WORKBOOK TO ACCOMPANY

FRENCH • VIERCK • FOSTER
ENGINEERING DRAWING AND GRAPHIC TECHNOLOGY

FOURTEENTH EDITION

Engineering Drawing and Graphic Technology Problems Book III

HUGH F. ROGERS
The Pennsylvania State University

McGRAW-HILL, INC.
New York   St. Louis   San Francisco   Auckland   Bogotá   Caracas   Lisbon
London   Madrid   Mexico   Milan   Montreal   New Delhi   Paris
San Juan   Singapore   Sydney   Tokyo   Toronto

**WORKBOOK TO ACCOMPANY FRENCH • VIERCK • FOSTER:
ENGINEERING DRAWING AND GRAPHIC TECHNOLOGY
Engineering Drawing and Graphic Technology Problems Book III**

Copyright © 1993, 1988 by McGraw-Hill, Inc. All rights reserved.
Printed in the United States of America. Except as permitted under the United States
Copyright Act of 1976, no part of this publication may be reproduced or distributed
in any form or by any means, or stored in a data base or retrieval system,
without the prior written permission of the publisher.

1 2 3 4 5 6 7 8 9 0 SEM SEM 9 0 9 8 7 6 5 4 3 2

P/N 053541-8
Part of
ISBN 0-07-911368-0

The editor was B. J. Clark;
the production supervisor was Leroy A. Young.
Semline, Inc., was printer and binder.

CONTENTS

1 and 2	Lettering	89	Skewed Lines
3 to 5	Architect's, Engineer's, and Metric Scales	90	Perpendiculars
6 to 14	Drawing Geometry	91	Piercing Points
15 to 19	Orthographic Sketching	92	Angles between Lines
20 to 26	Orthographic Drawing	93	Angles Between Planes
27 to 28	Inversion Problems (Orthographic)	94	Angles Between Pines and Planes
29 to 32	Orthographic Drawing	95	Strike and Dip
33 to 37	Isometric Sketching	96	Revolution
38 to 41	Isometric Drawing	97	Space Geometry Problem
42 to 48	Oblique Sketching	98 to 102	Intersections
49	Isometric and Oblique Drawing	103 to 107	Developments
50 to 60	Sections	108 to 112	Working Drawings
61 to 68	Dimensioning	113 to 116	Graphs
69	Tolerances	117	Graphical Differentiation
70	Metric Tolerances	118	Graphical Integration
71 to 73	Geometric Tolerancy	119 to 121	Vectors
74	Points and Lines—Space Geometry	122	Threads
75	Lines—Space Geometry	123	Metric Threads
76	Planes—Space Geometry	124	Spring and Key Drawing
77	Bearing and Slope of a Line	125	Gear Drawing
78	Bearing, True Length and Grade	126	Plate Cam Drawing
79	Bearing, True Length and Slope	127 and 128	Welding Symbols
80	True Lengths Problems	129	Welding Drawings
81	Point View of Line; Edge View of Plane	130 and 131	Electronic Drawing
82	True Size of Planes	132	Pipe Drawing
83	Slope of Planes	133	Structural Drawing
84 to 87	Auxillary Views	134	Topographic Drawing
88	Shortest Lines		

SEVERAL OF THE DRAWING PROBLEMS IN THIS WORKBOOK MAKE REFERENCE TO OPERATIONS OR "CALLOUTS" FOR VARIOUS TYPES OF HOLES. DRAWN BELOW ARE ILLUSTRATIONS FOR A FEW OF THE TYPES OF HOLES SHOWN IN ORTHOGRAPHIC STYLE. AN ACCOMPANYING NOTE FOR EACH HOLE IS GIVEN FOR YOUR UNDERSTANDING OF THE ORDER OF HOLE FORMATION.

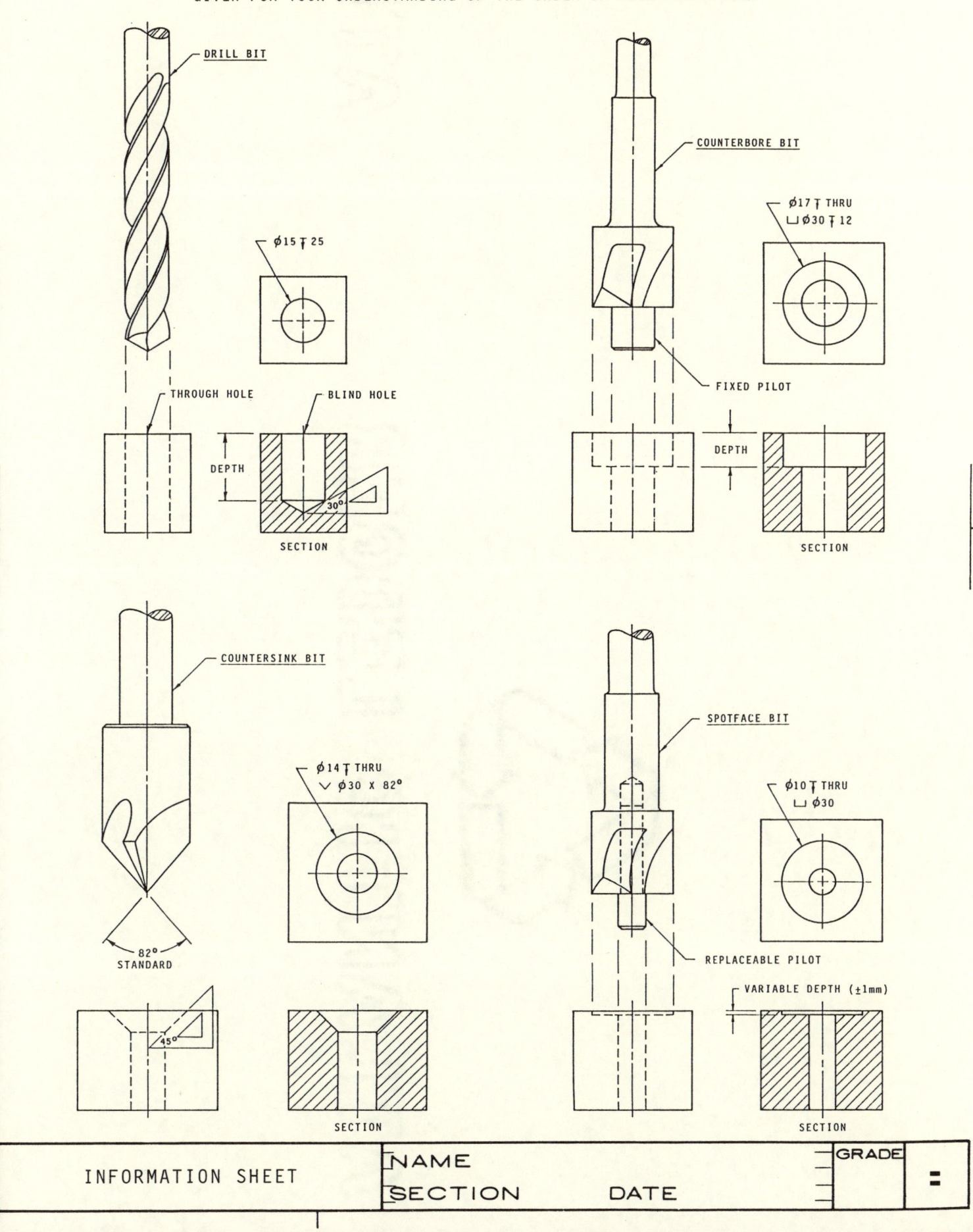

INFORMATION SHEET

1 REPEAT EACH LETTER AND NUMBER USING THE GUIDELINES PROVIDED. USE A SOFT LEAD (F OR HB) FOR LETTERING. A CONICAL POINT WITH A SLIGHTLY ROUNDED TIP IS THE MOST PREFERRED SHAPE OF POINT. ROTATE THE PENCIL BETWEEN THE THUMB AND INDEX FINGER AFTER EVERY FEW STROKES TO KEEP THE STROKES UNIFORM.

A V K X Y N
Z L T F E H
J U D P R B
C G Q S M W

NOTE: THE LETTER I AND THE NUMBER 1 ARE THE SAME.   I

2   3   4
5   6   7
8   9   0

2 LETTER THESE WORDS, COMMON TO ENGINEERING, IN THE GUIDELINES BELOW.

BROACH   ZONE   VALVE   KEY   FILLET   LIMIT   TORQUE   COPE   HUB

SECTION   METAL   BORE   HEX   SQUARE   BEARING   WORK   ANALYZE

3 OBSERVE THE FORM AND PROPORTION OF THE ABOVE LETTERS AND WORDS AND REPEAT THE FOLLOWING SENTENCES IN THE GUIDELINES PROVIDED BELOW.

LETTERING ON AN ENGINEERING DRAWING OR SKETCH SHOULD CONVEY THE REQUIRED INFORMATION CLEARLY AND EFFICIENTLY. THE BASIS OF GOOD LETTERING IS THE SINGLE STROKE GOTHIC ALPHABET.

LETTERING	NAME		GRADE
	SECTION	DATE	1

1 REPEAT EACH LETTER AND NUMBER USING THE GUIDELINES PROVIDED. USE A SOFT LEAD (F OR HB) FOR LETTERING. A CONICAL POINT WITH A SLIGHTLY ROUNDED TIP IS THE MOST PREFERRED SHAPE OF POINT. ROTATE THE PENCIL BETWEEN THE THUMB AND INDEX FINGER AFTER EVERY FEW STROKES TO KEEP THE STROKES UNIFORM.

A V K X Y N
Z L T F E H
J U D P R B
Q C G S M W

NOTE: THE LETTER I AND THE NUMBER 1 ARE THE SAME. *I*

2   3   4
5   6   7
8   9   0

2 LETTER THESE WORDS, COMMON TO ENGINEERING, IN THE GUIDELINES BELOW.

TAPER PIN DEEP SPOTFACE REAM GRIND KNURL CHAMFER RIB

▶ _____
▶ _____

DRILL GAUGE BOLT KEYSEAT PRESS DRAG SPOKE TREATMENT

▶ _____
▶ _____

3 OBSERVE THE FORM AND PROPORTION OF THE ABOVE LETTERS AND WORDS AND REPEAT THE FOLLOWING SENTENCES IN THE GUIDELINES PROVIDED BELOW.

LETTERING ON AN ENGINEERING DRAWING OR SKETCH SHOULD CONVEY THE REQUIRED INFORMATION CLEARLY AND EFFICIENTLY. THE BASIS OF GOOD LETTERING IS THE SINGLE STROKE GOTHIC ALPHABET.

▷ _____
▷ _____
▷ _____
▷ _____

LETTERING	NAME	GRADE
	SECTION DATE	2

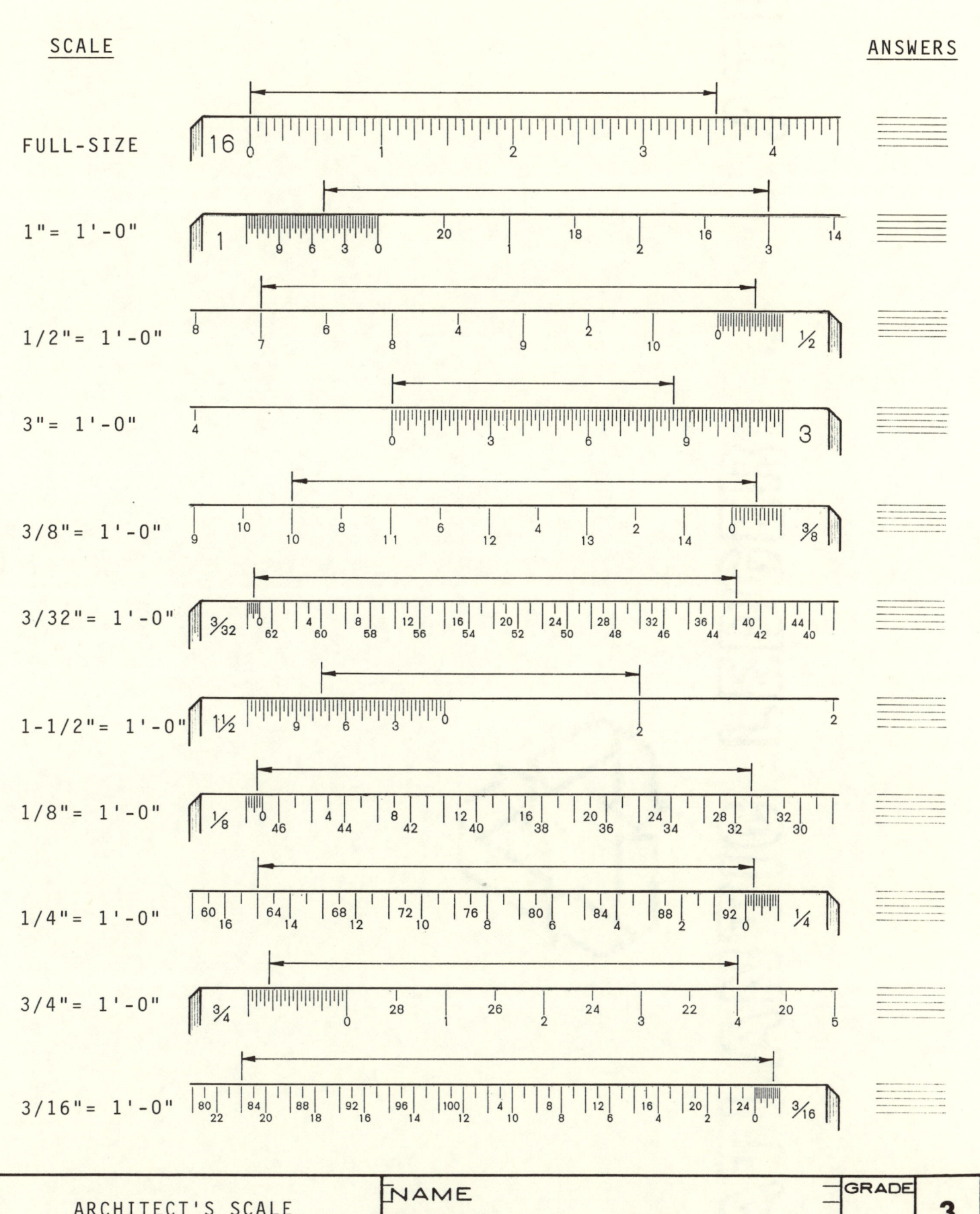

USING THE SCALES SHOWN BELOW, DETERMINE THE LENGTHS DEFINED.
LETTER THE ANSWERS WITHIN THE GUIDELINES PROVIDED.

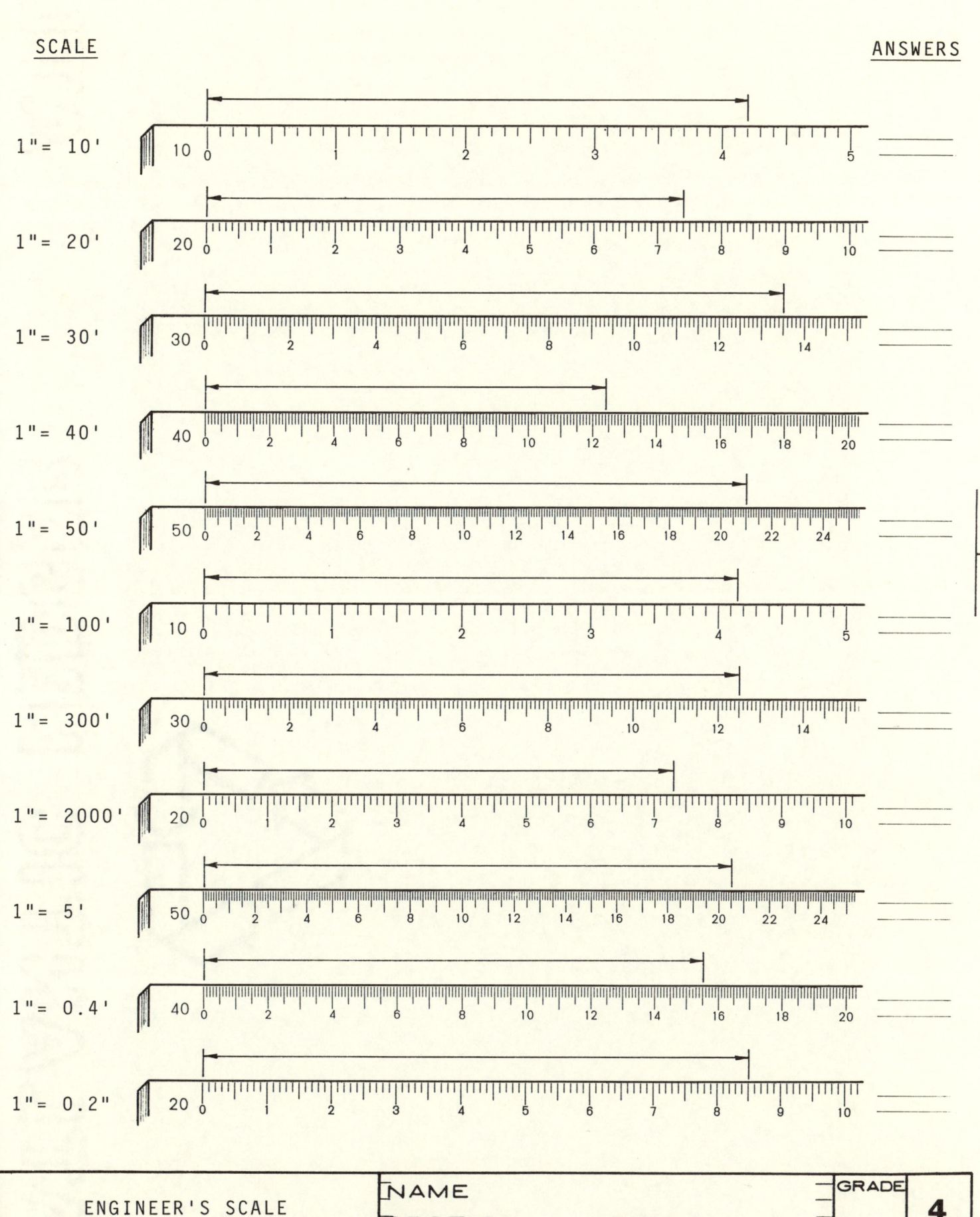

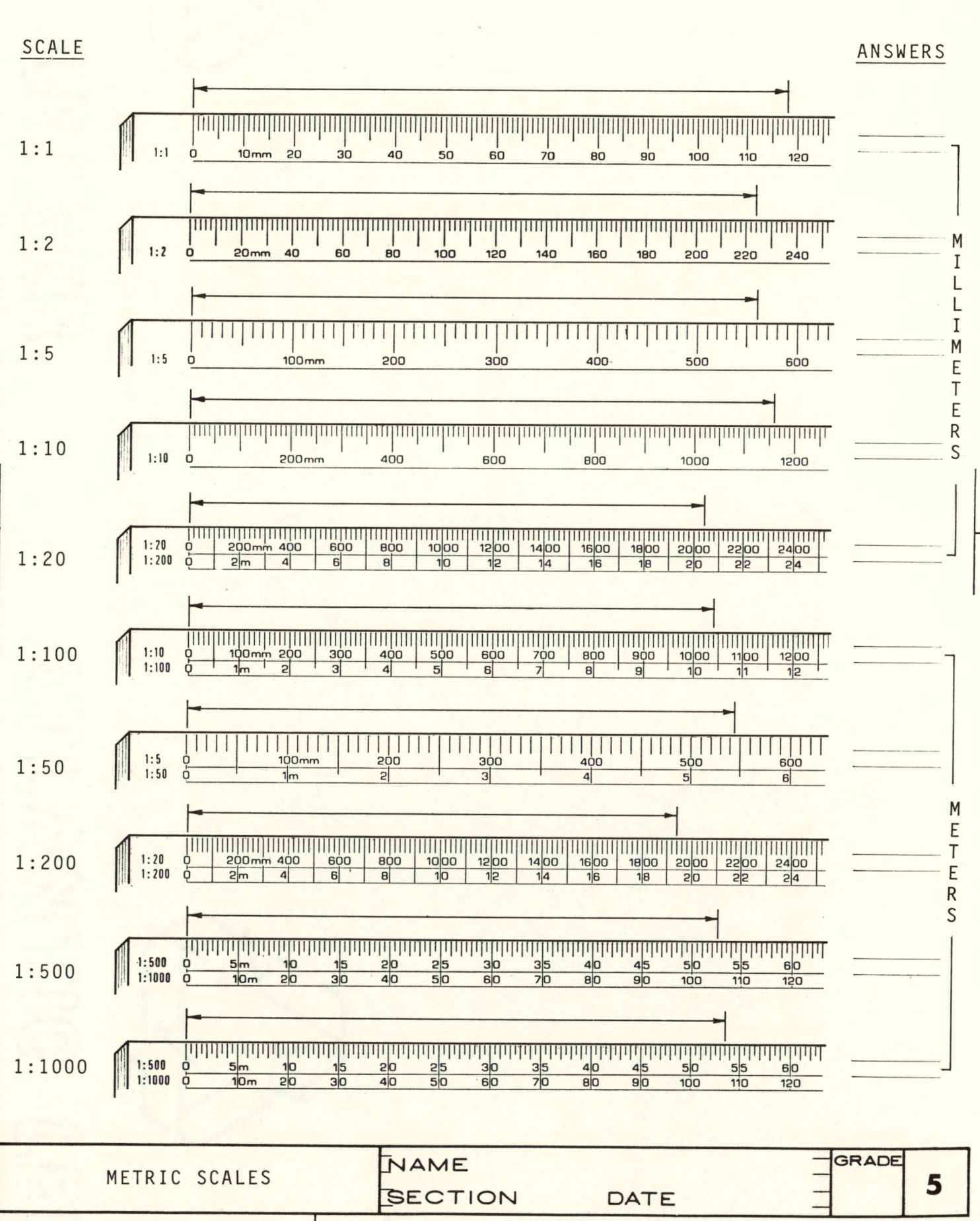

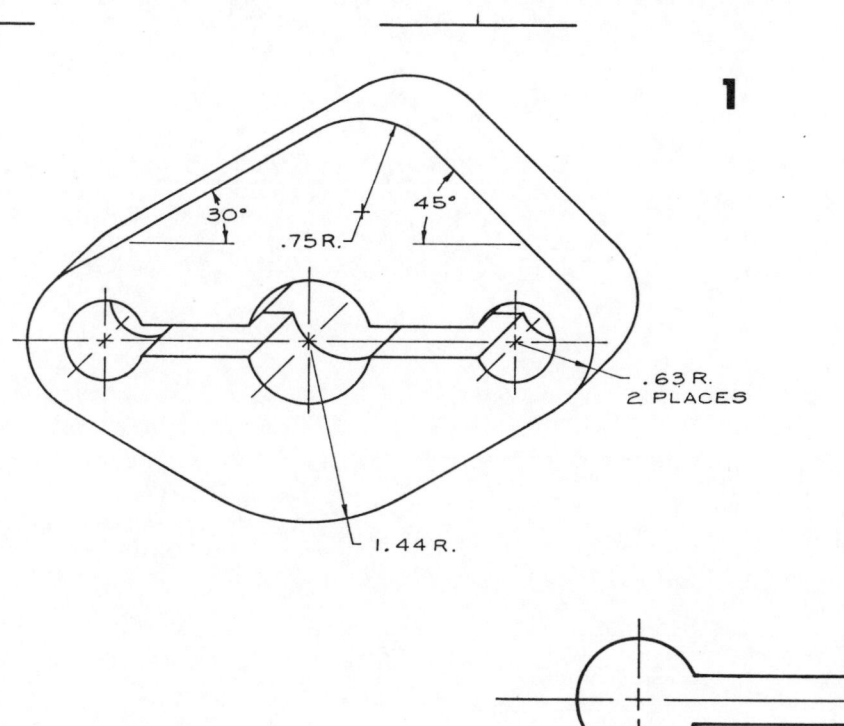

1

COMPLETE THE FULL-SIZE DRAWING OF THIS PROBLEM BY USING THE DIMENSIONS GIVEN IN THE SMALL PICTORIAL DRAWING. MARK POINTS OF TANGENCY. SHOW CONSTRUCTION.

DRAWING SCALE: 1" = 1"

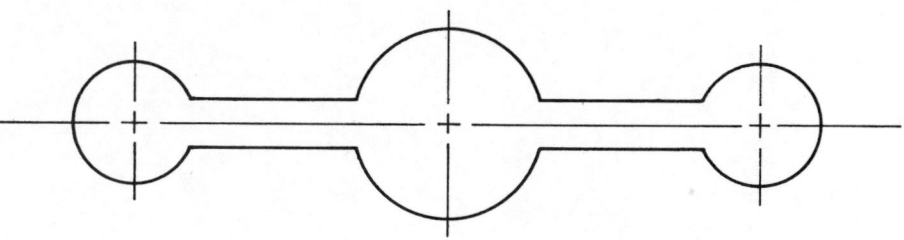

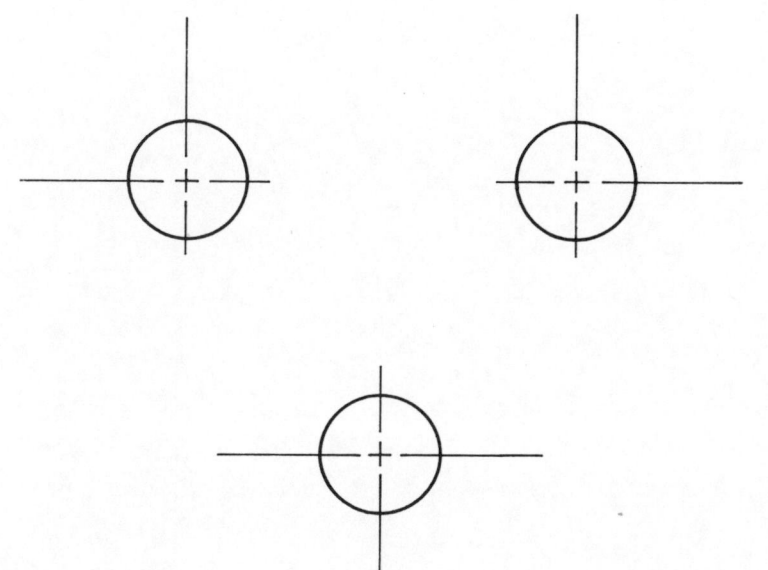

COMPLETE THE FULL-SIZE DRAWING OF THIS PROBLEM BY USING THE DIMENSIONS GIVEN IN THE SMALL PICTORIAL DRAWING. MARK POINTS OF TANGENCY. SHOW CONSTRUCTION.

DRAWING SCALE: 1" = 1"

2

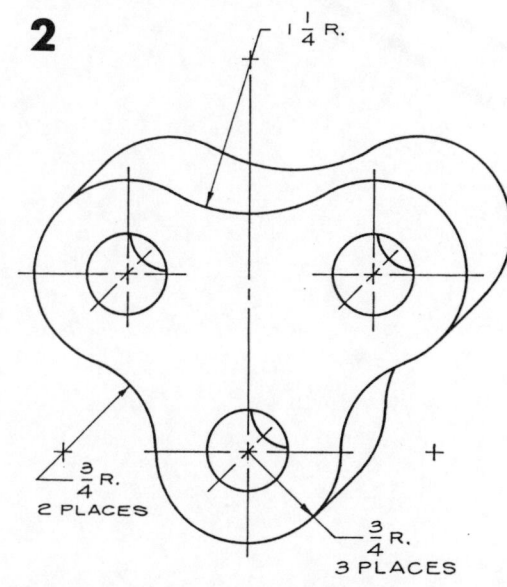

| DRAWING GEOMETRY | NAME | GRADE | 6 |
| | SECTION   DATE | | |

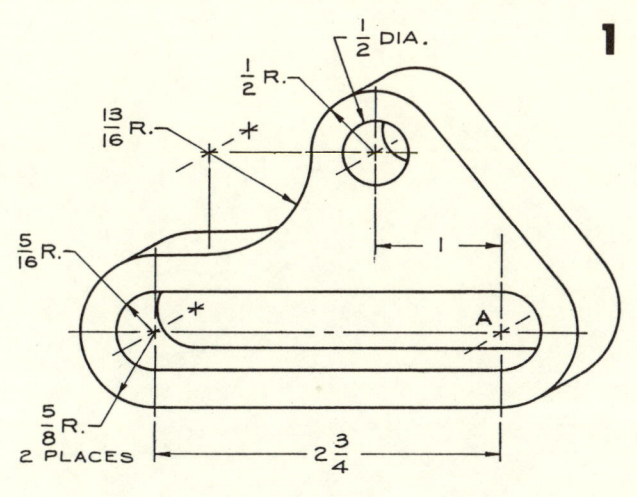

1 COMPLETE THE FULL-SIZE DRAWING OF THIS PROBLEM BY USING THE DIMENSIONS GIVEN IN THE SMALL PICTORIAL DRAWING. MARK POINTS OF TANGENCY. SHOW CONSTRUCTION.

DRAWING SCALE: 1" = 1"

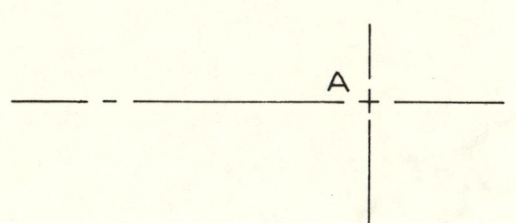

COMPLETE THE FULL-SIZE DRAWING OF THIS PROBLEM BY USING THE DIMENSIONS GIVEN IN THE SMALL PICTORIAL DRAWING. MARK POINTS OF TANGENCY. SHOW CONSTRUCTION.

DRAWING SCALE: 1" = 1"

2

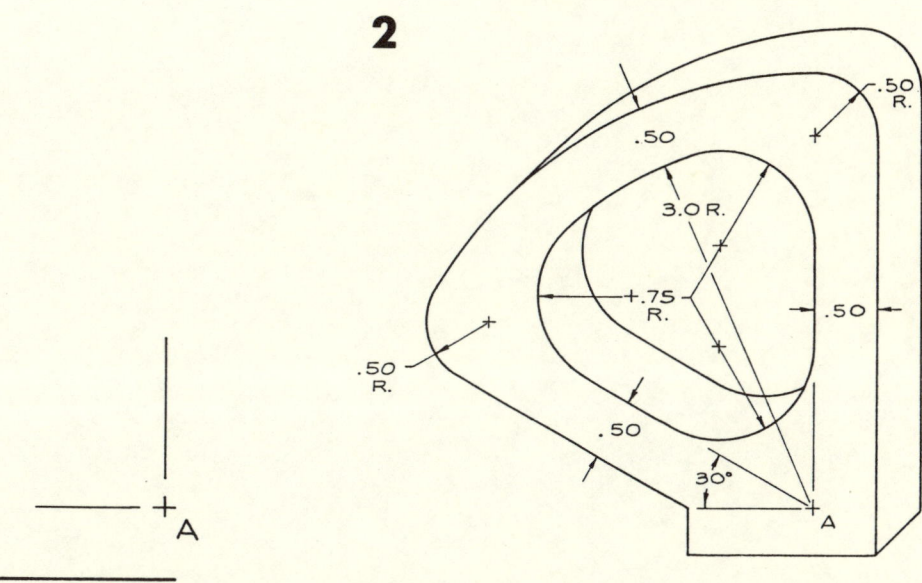

DRAWING GEOMETRY | NAME | SECTION | DATE | GRADE | 7

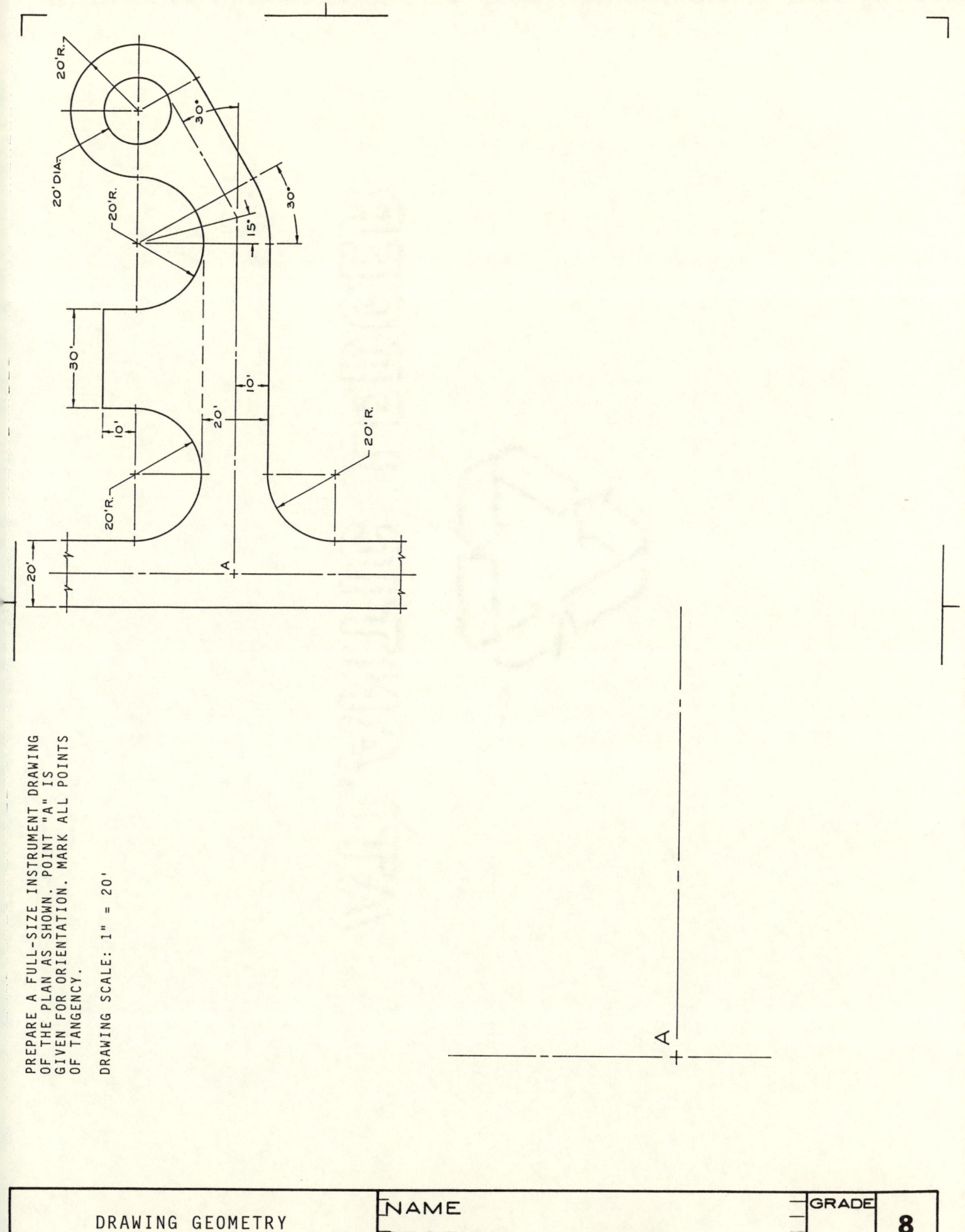

PREPARE A FULL-SIZE DRAWING OF THE PROBLEM AS SHOWN IN THE SMALL DRAWING.

DRAWING SCALE: 1" = 1"

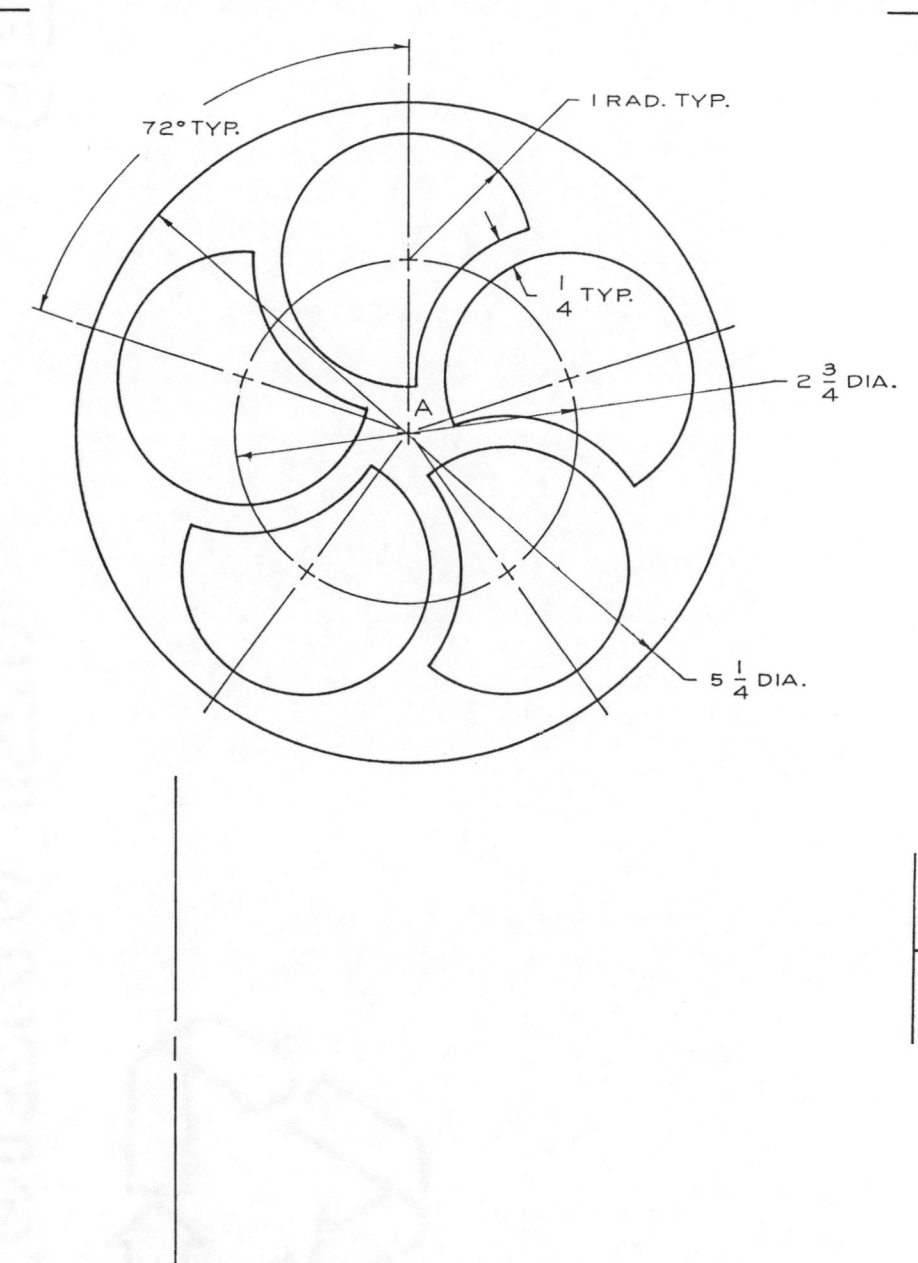

DRAWING GEOMETRY

NAME
SECTION   DATE

GRADE
9

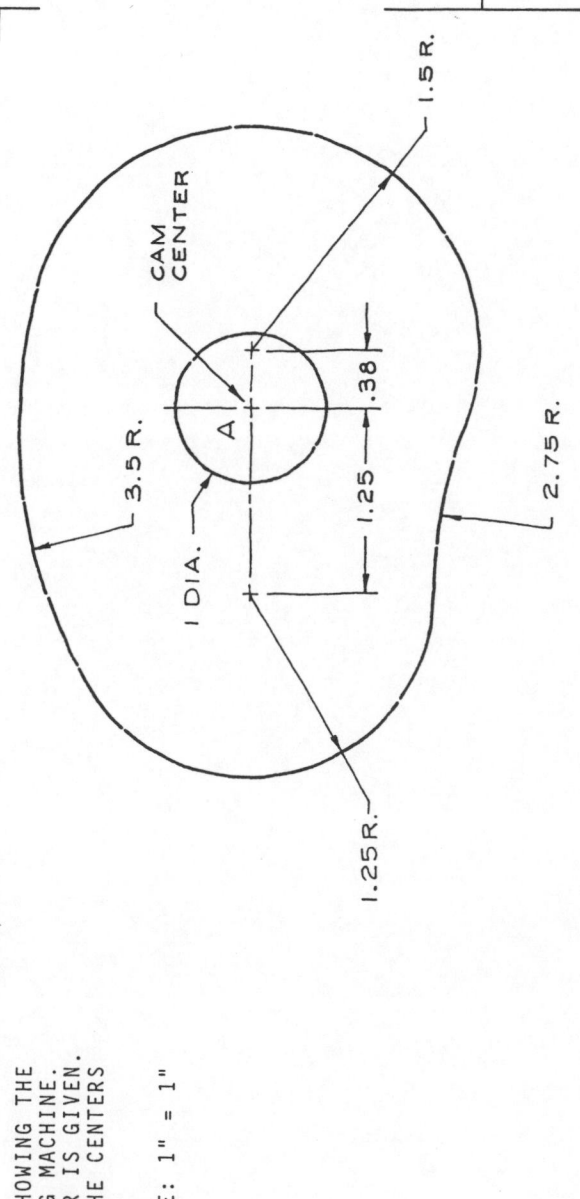

AT THE RIGHT IS AN ENGINEER'S SKETCH SHOWING THE PROPOSED PROFILE OF A CAM FOR A SHAPING MACHINE. DRAW THE CAM FULL-SIZE. THE CAM CENTER IS GIVEN. SHOW THE CONSTRUCTION USED TO OBTAIN THE CENTERS OF RADII.

MARK ALL TANGENT POINTS   DRAWING SCALE: 1" = 1"

DRAWING GEOMETRY

NAME
SECTION   DATE

GRADE
10

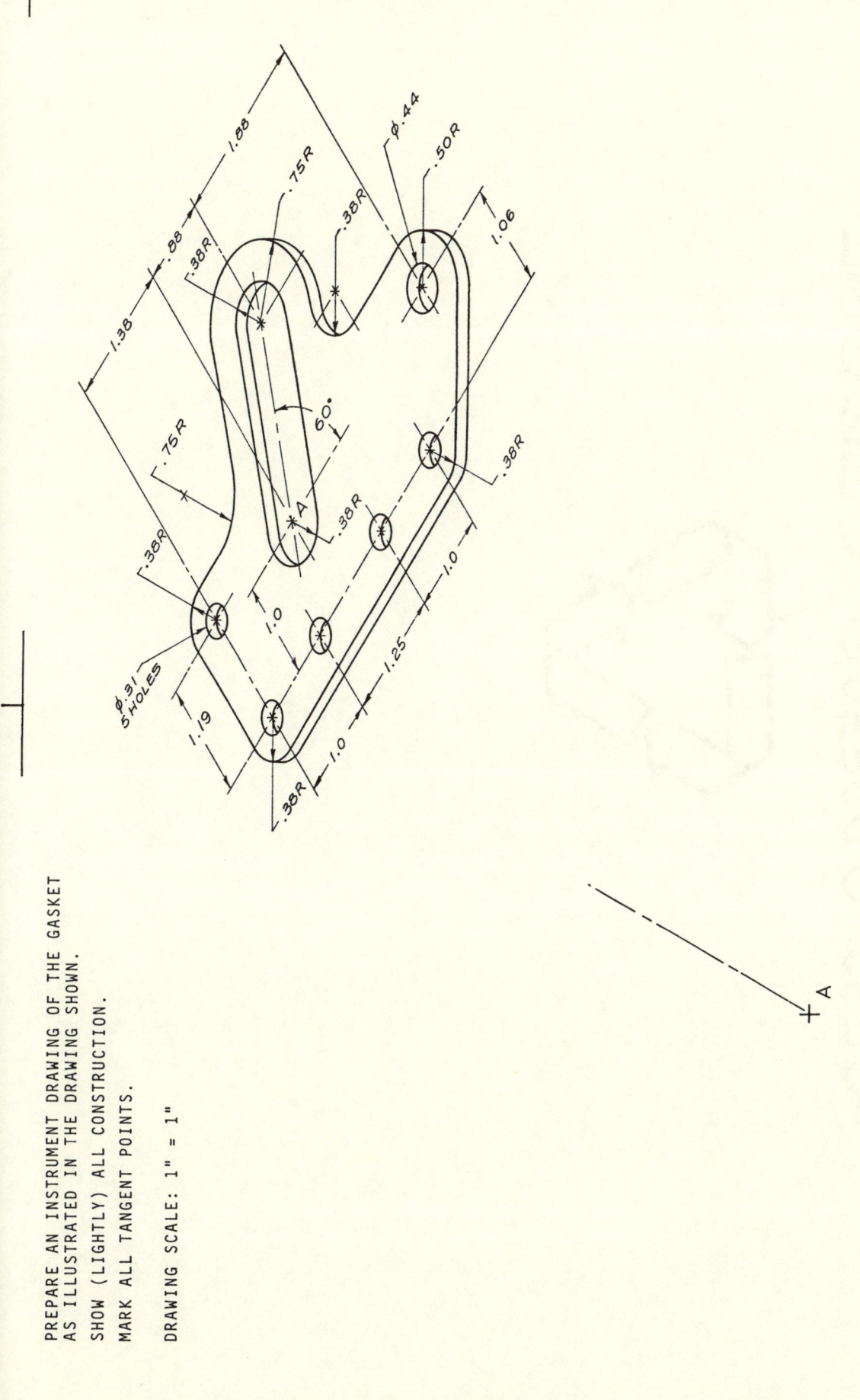

PREPARE AN INSTRUMENT DRAWING OF THE GASKET AS ILLUSTRATED IN THE DRAWING SHOWN.
SHOW (LIGHTLY) ALL CONSTRUCTION.
MARK ALL TANGENT POINTS.

DRAWING SCALE: 1" = 1"

DRAWING GEOMETRY	NAME	GRADE	11
	SECTION DATE		

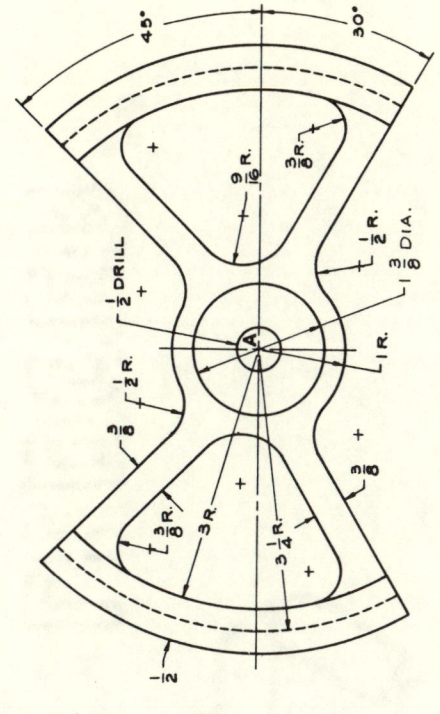

BEGINNING WITH THE REFERENCE POINT A, PREPARE A FULL SIZE
DRAWING OF THE RUDDER QUADRANT USING THE DIMENSIONS GIVEN
IN THE EXAMPLE DRAWING.

SHOW CLEARLY ALL CENTERS OF RADII AND POINTS OF TANGENCY.

DRAWING SCALE: 1"= 1"

DRAWING GEOMETRY	NAME	GRADE	12
	SECTION DATE		

BEGINNING WITH THE REFERENCE POINT A, PREPARE A FULL SIZE DRAWING OF THE HINGE FITTING USING THE DIMENSIONS GIVEN IN THE EXAMPLE DRAWING.

SHOW CLEARLY ALL CENTERS OF RADII AND POINTS OF TANGENCY.

DRAWING SCALE: 1" = 1"

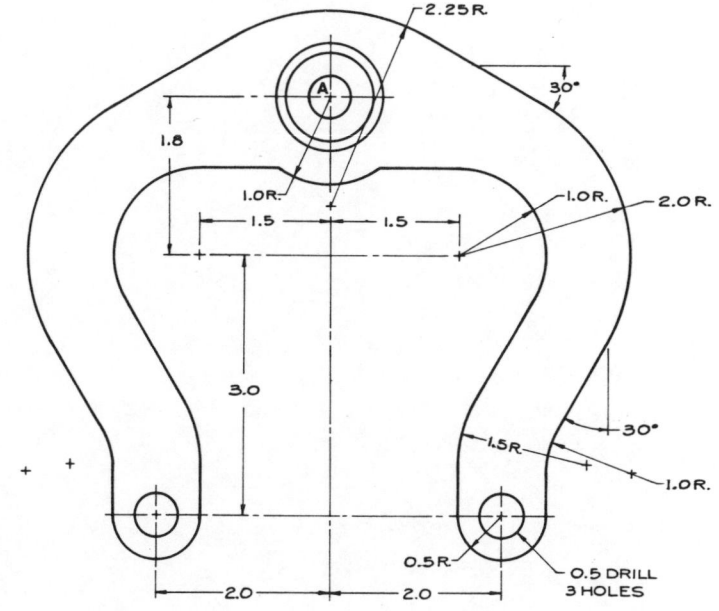

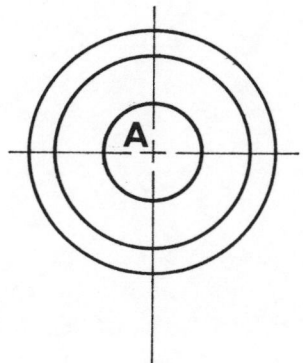

DRAWING GEOMETRY

NAME
SECTION   DATE

GRADE

13

PREPARE A FULL-SIZE DRAWING OF THE PROBLEM
AS SHOWN IN THE SMALL DRAWING TO THE RIGHT.

<u>MARK ALL TANGENT POINTS</u>

DRAWING SCALE: 1"= 1"

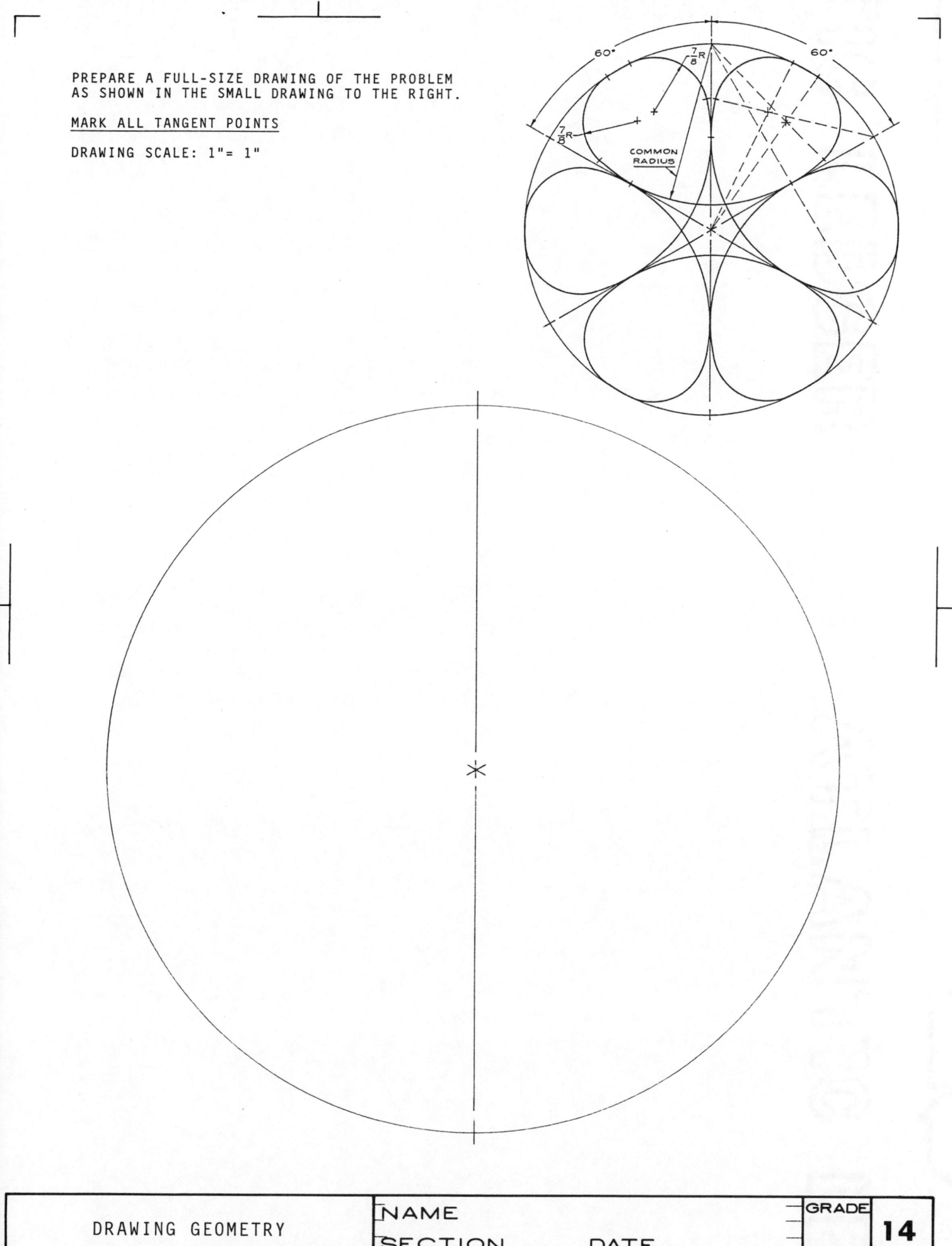

DRAWING GEOMETRY	NAME		GRADE	14
	SECTION	DATE		

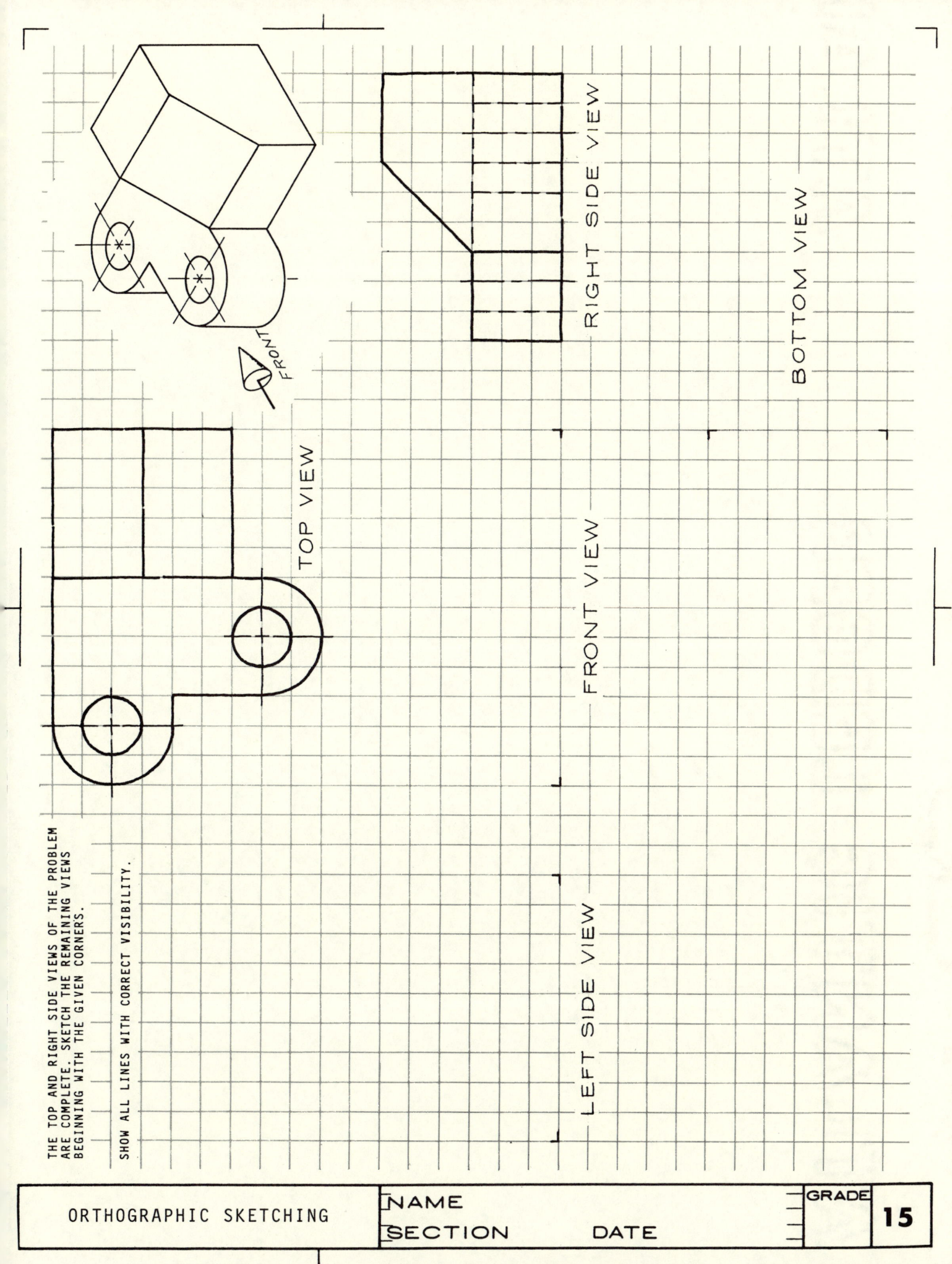

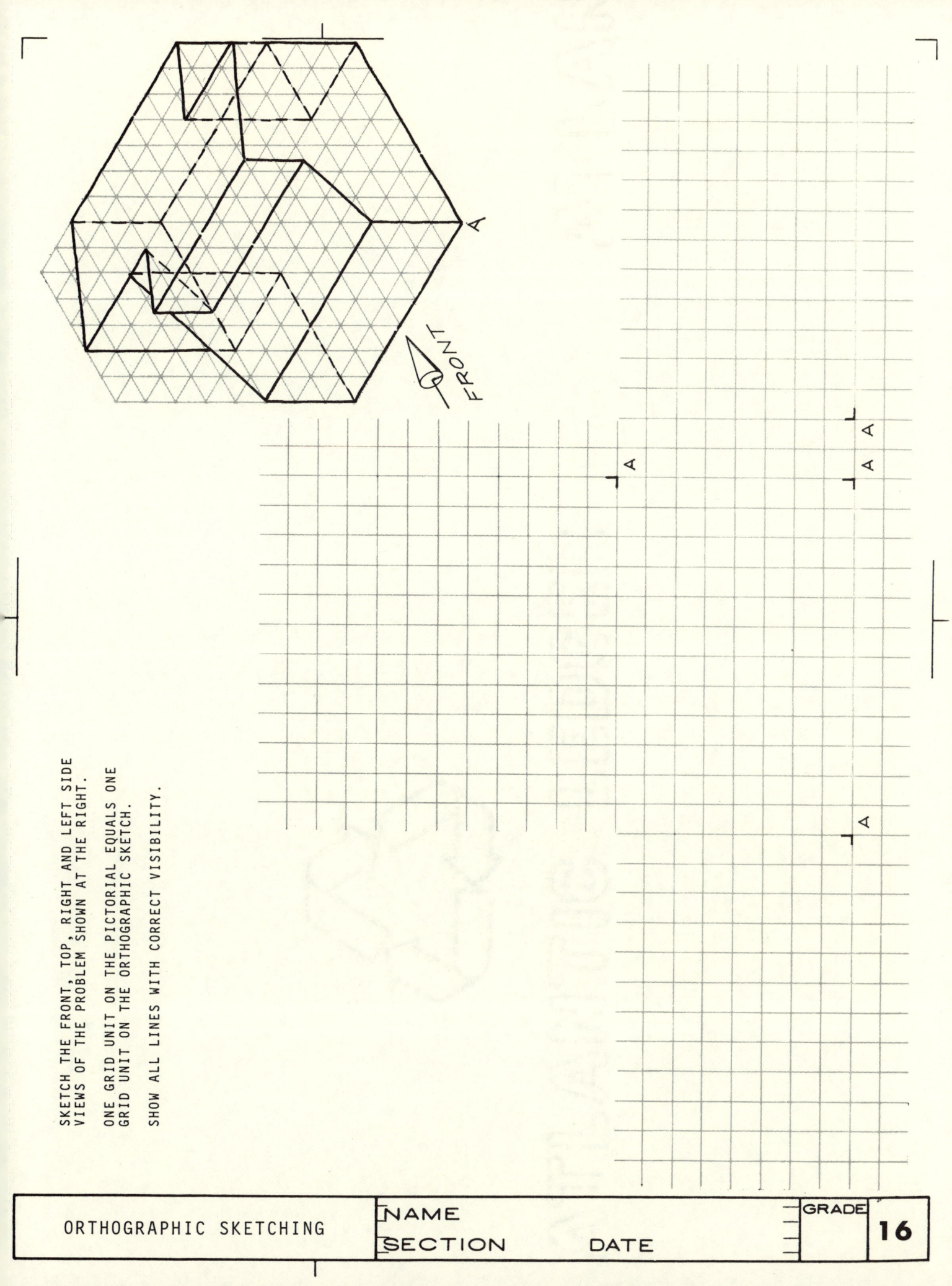

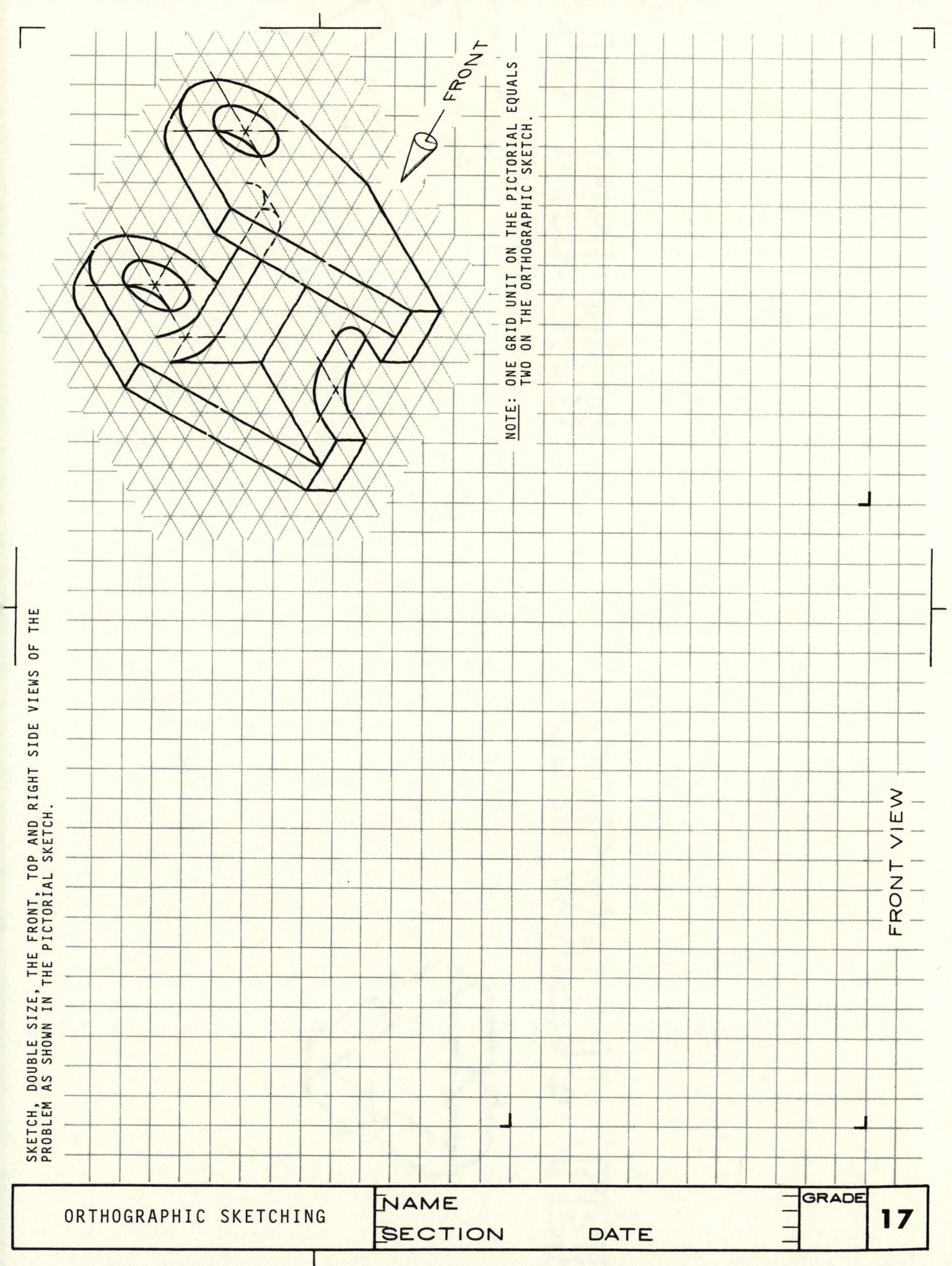

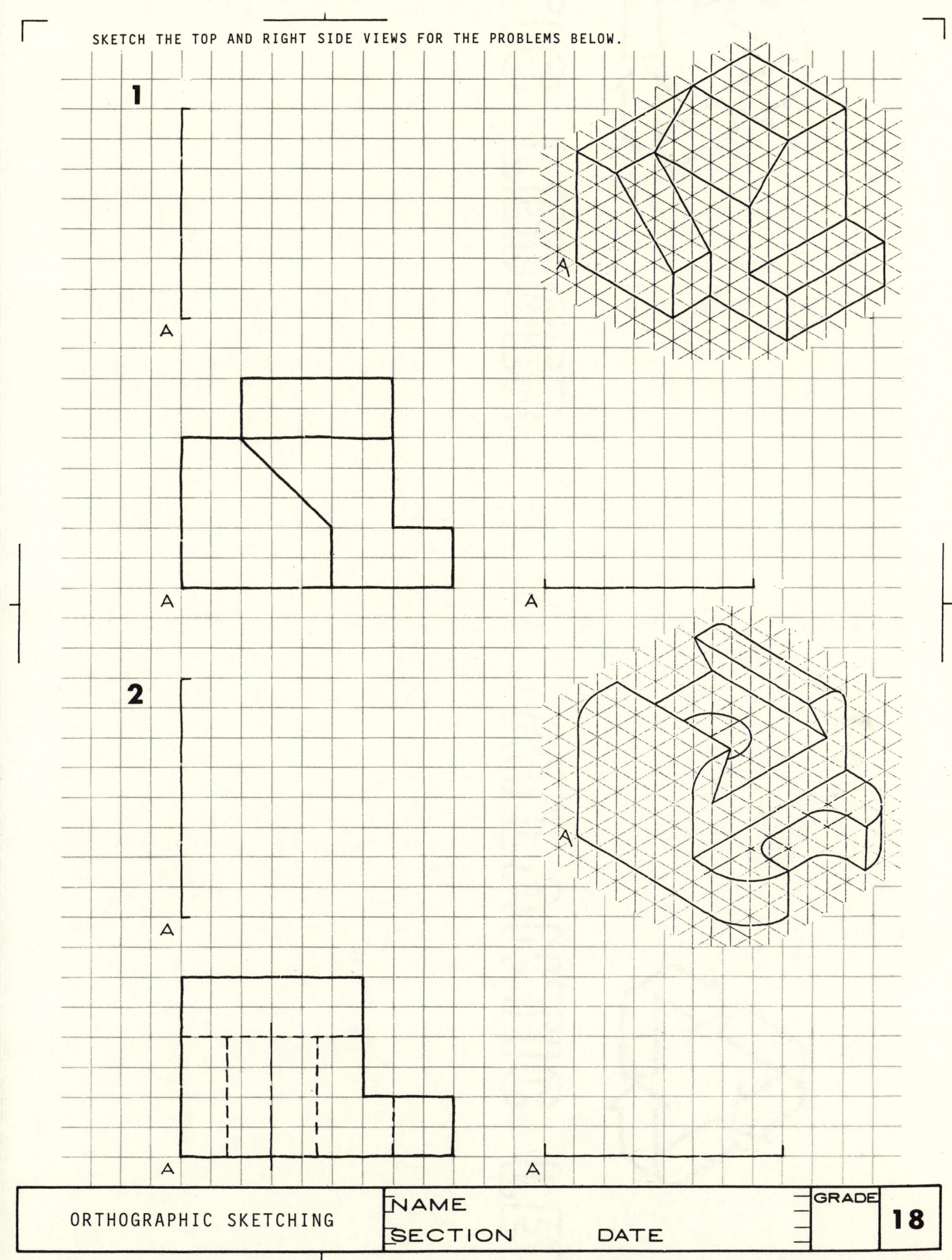

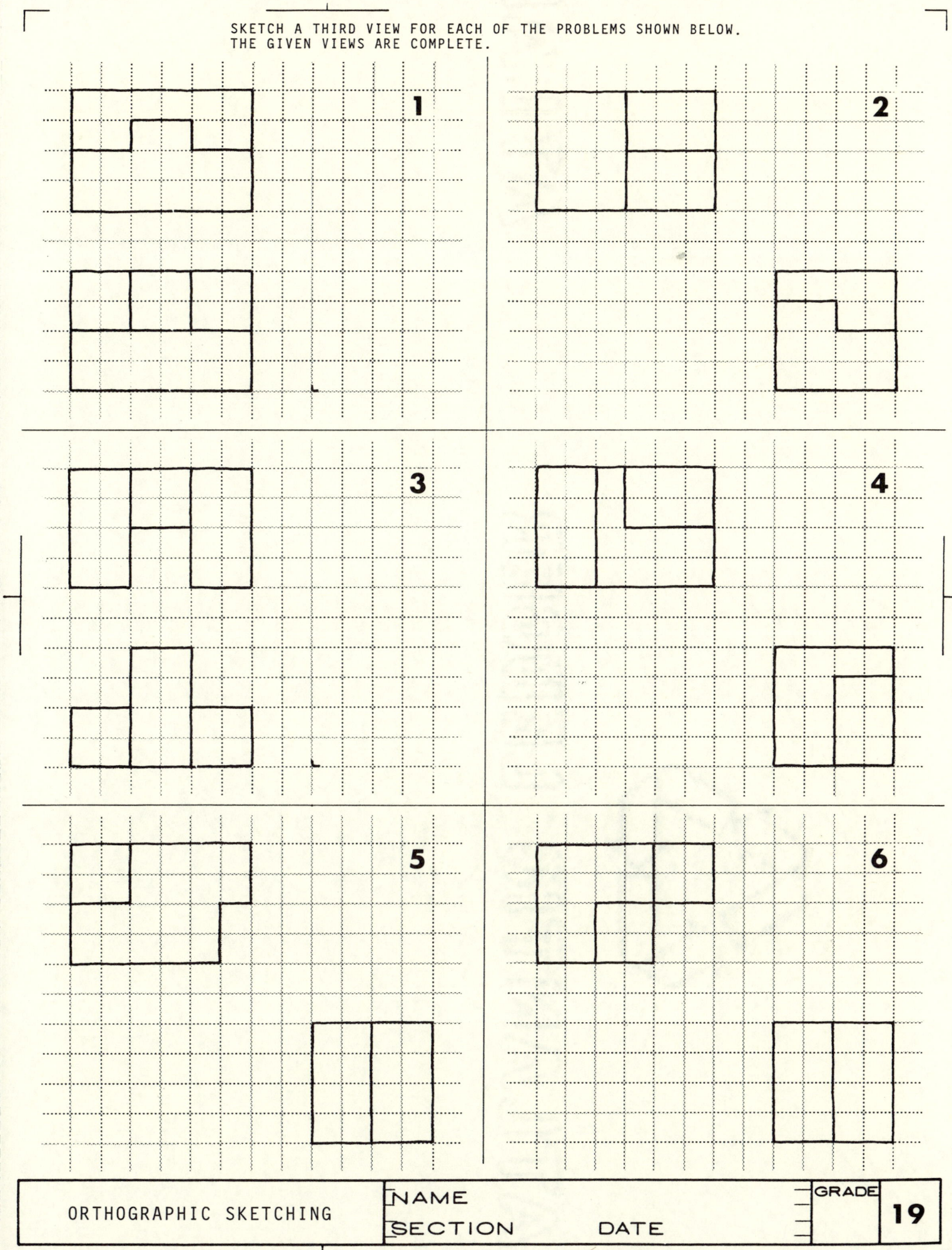

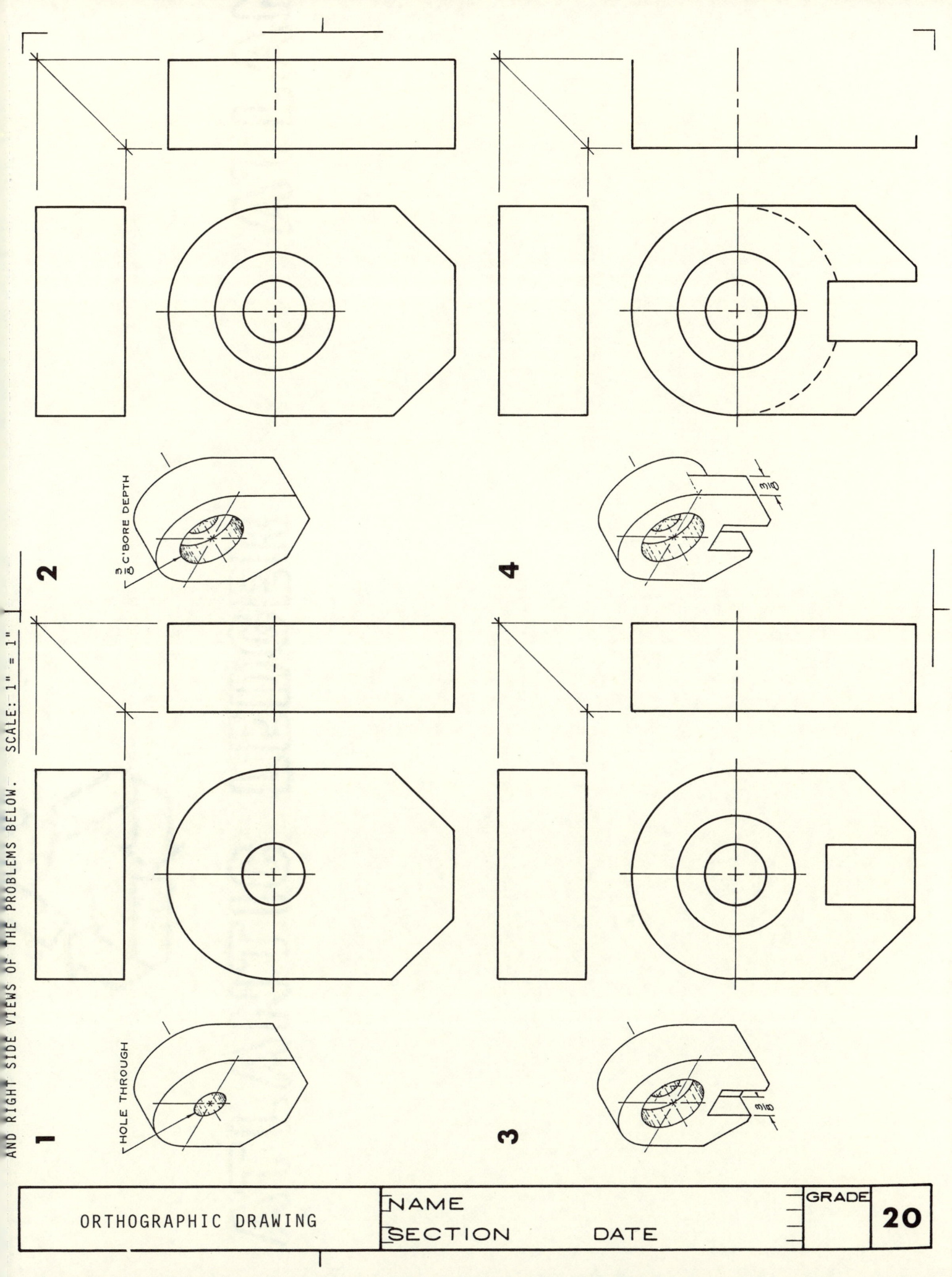

DRAW THE MISSING FRONT VIEW FOR EACH OF THE PROBLEMS BELOW.

1

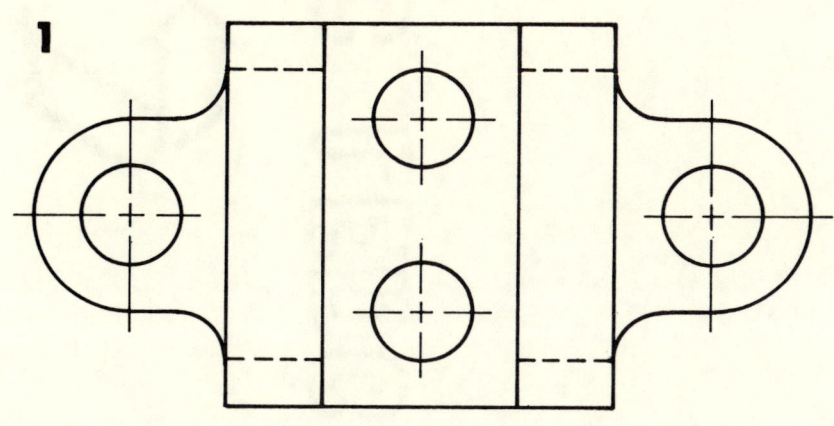
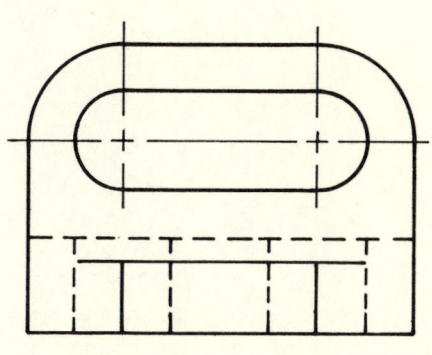

2

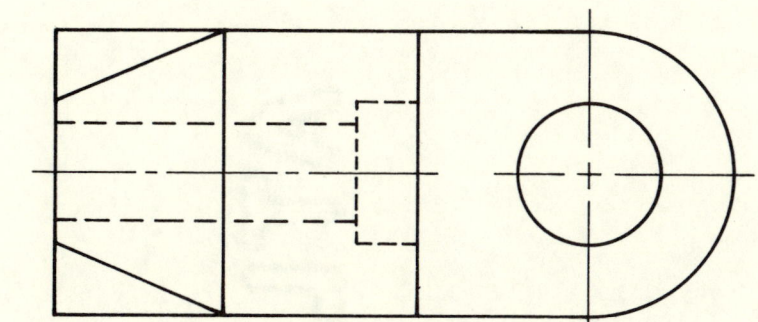
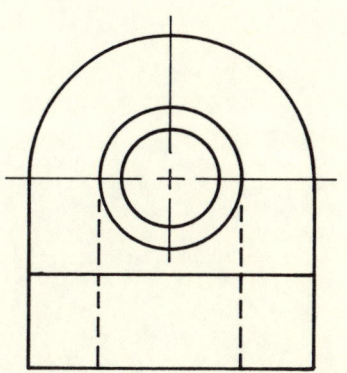

ORTHOGRAPHIC DRAWING	NAME	GRADE	21
	SECTION DATE		

USING INSTRUMENTS, COMPLETE THE VIEWS OF THE OBJECT BELOW.

ORTHOGRAPHIC DRAWING | NAME | SECTION | DATE | GRADE | 22

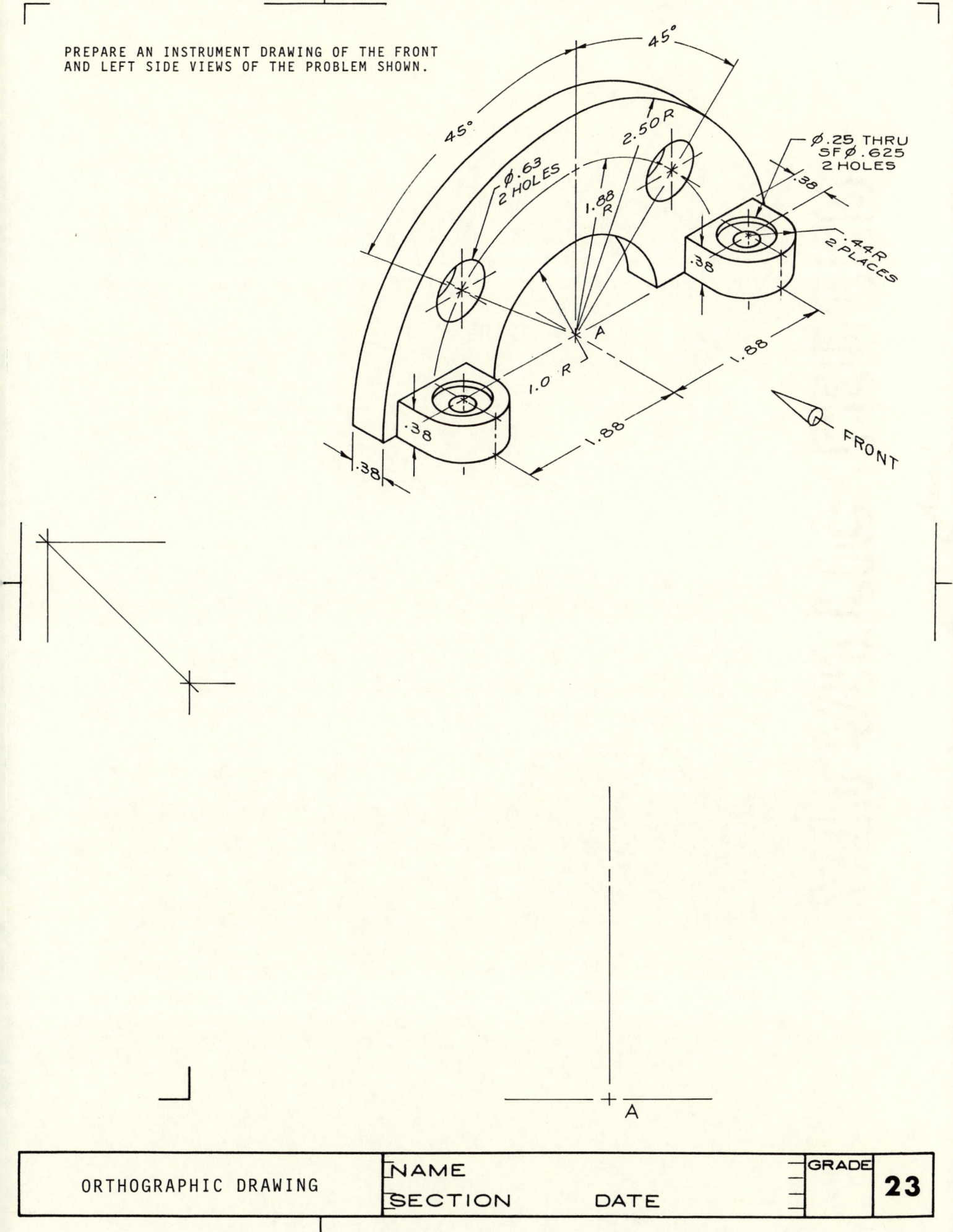

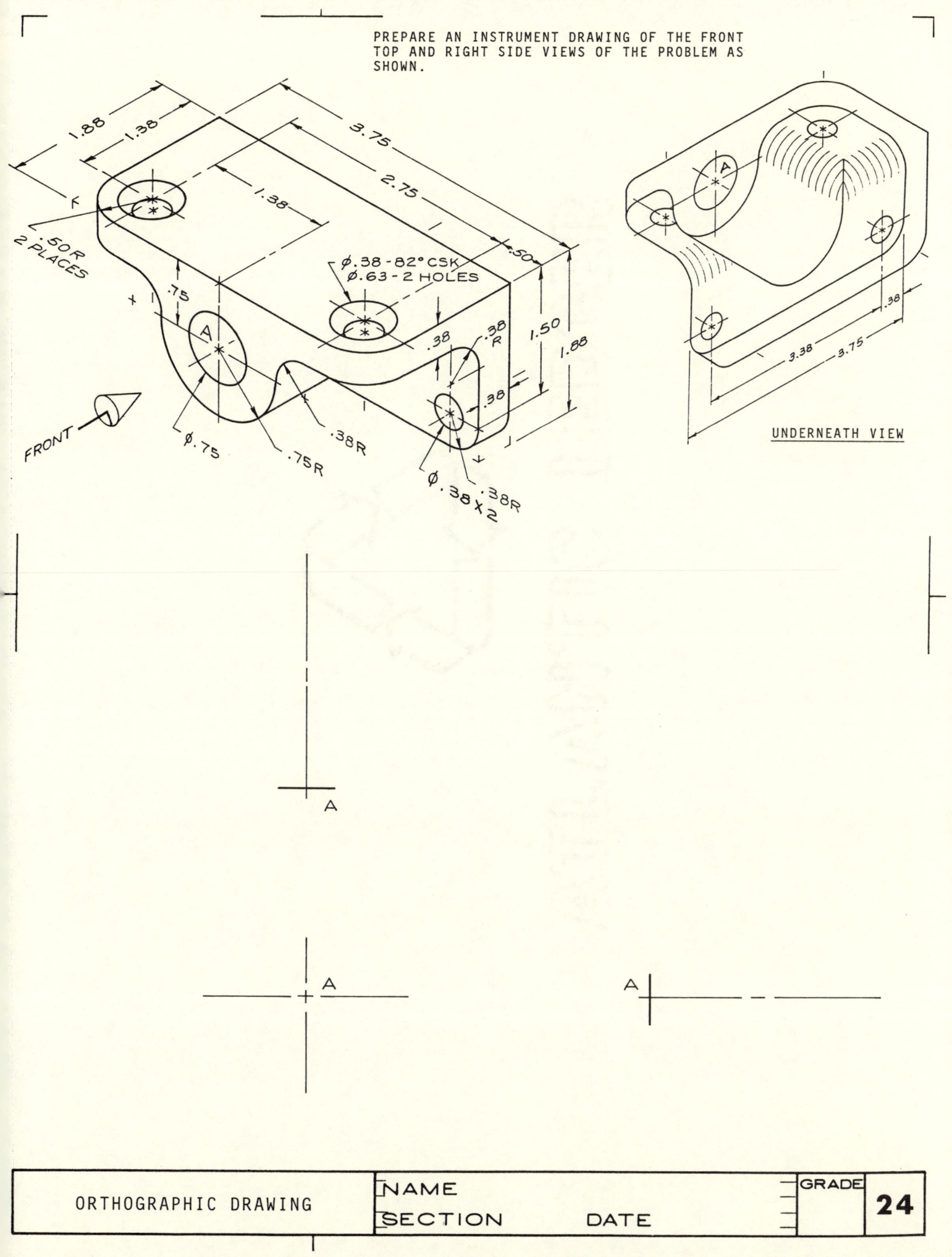

PREPARE AN INSTRUMENT DRAWING OF THE FRONT TOP AND RIGHT SIDE VIEWS OF THE PROBLEM AS SHOWN.

ORTHOGRAPHIC DRAWING	NAME	GRADE
	SECTION DATE	25

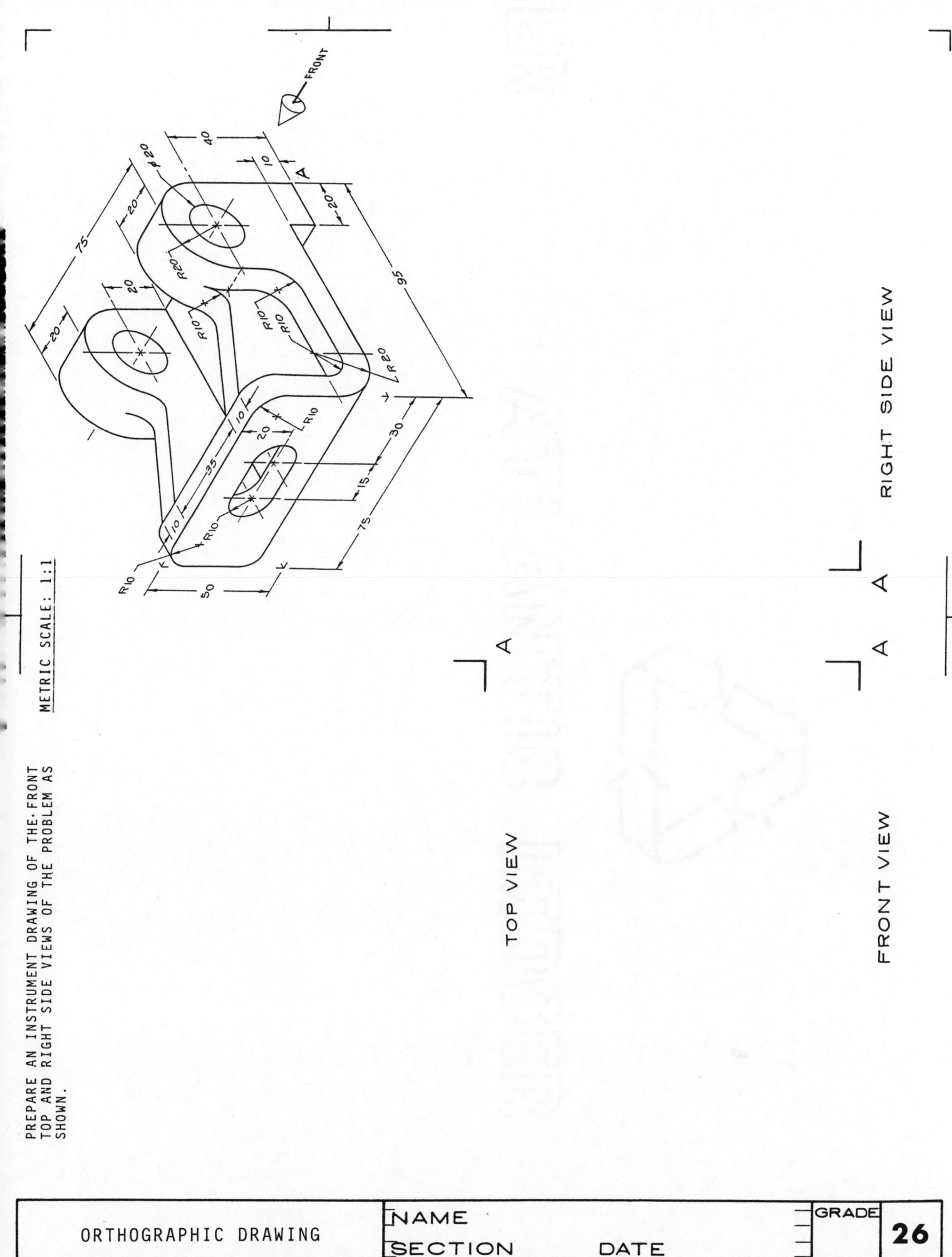

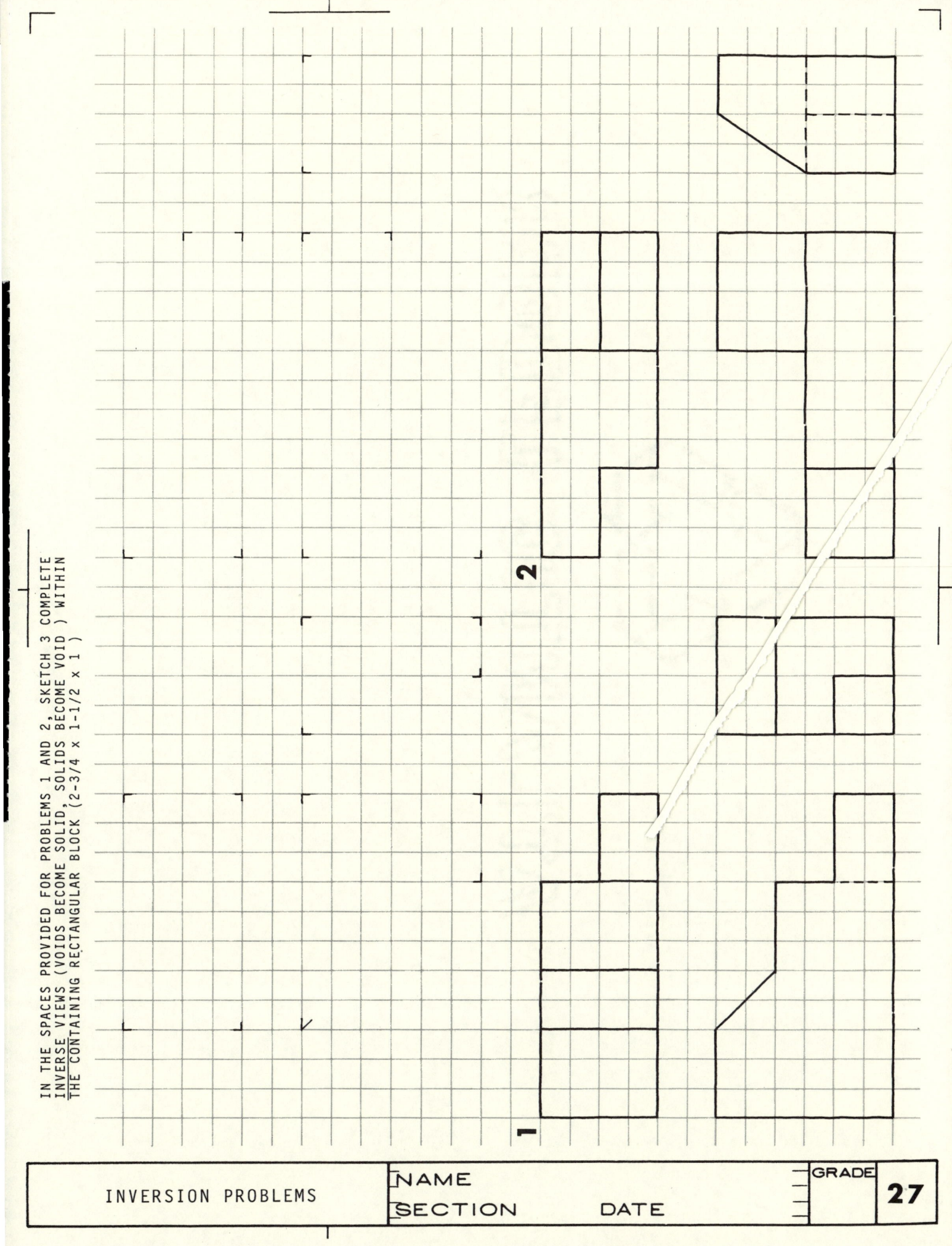

IN THE SPACES PROVIDED FOR PROBLEMS 1 AND 2, SKETCH 3 COMPLETE INVERSE VIEWS (VOIDS BECOME SOLID, SOLIDS BECOME VOID) WITHIN THE CONTAINING RECTANGULAR BLOCK (2-3/4 x 1-1/2 x 1)

INVERSION PROBLEMS

NAME
SECTION   DATE

GRADE
27

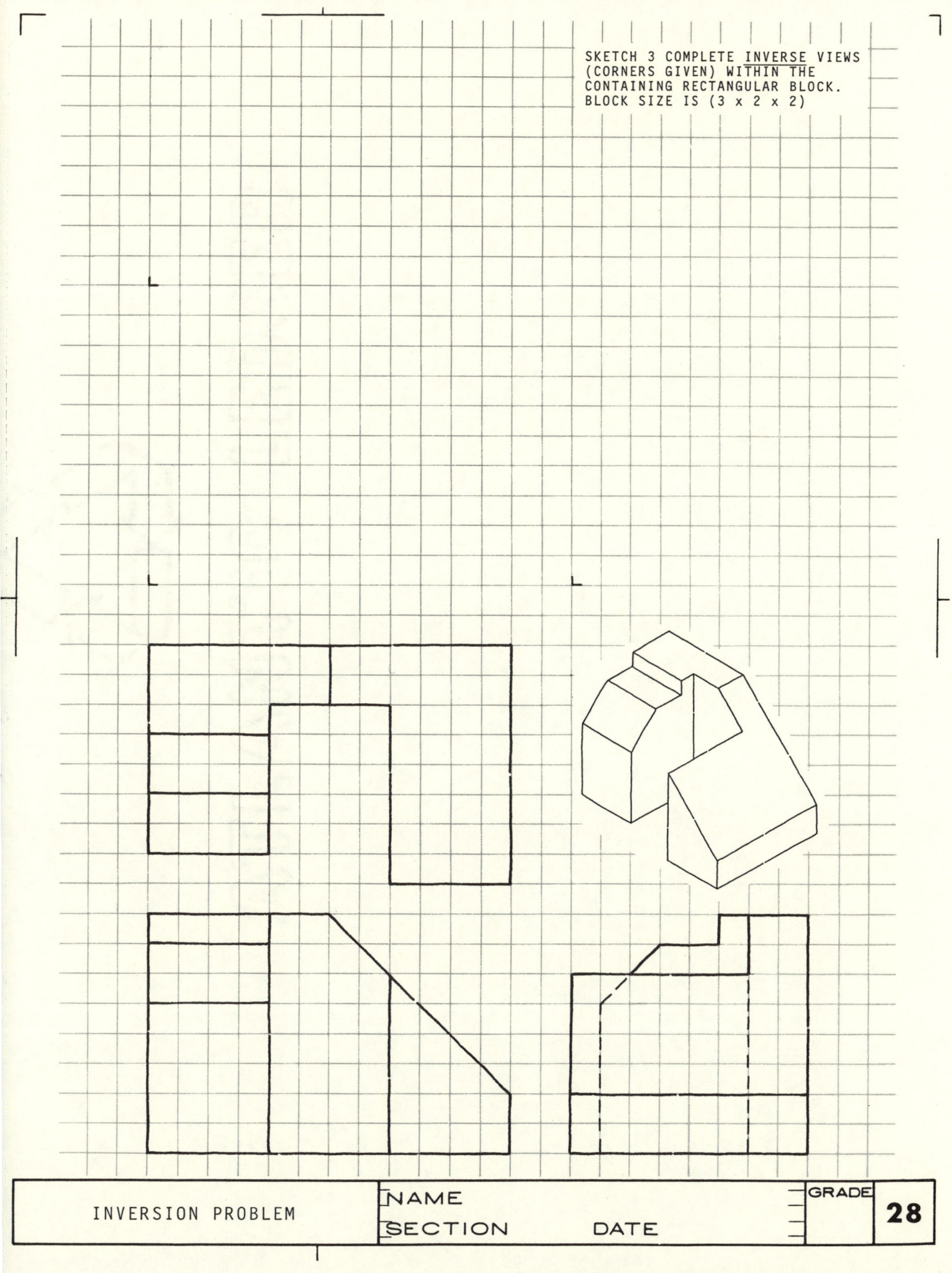

LOCATE AND DRAW THE MISSING LINES AND COMPLETE THE GIVEN VIEWS USING INSTRUMENTS.

| ORTHOGRAPHIC DRAWING | 29 |

LOCATE AND DRAW THE MISSING LINES AND COMPLETE THE GIVEN VIEWS USING INSTRUMENTS.

ORTHOGRAPHIC DRAWING	NAME — SECTION — DATE — GRADE — 30

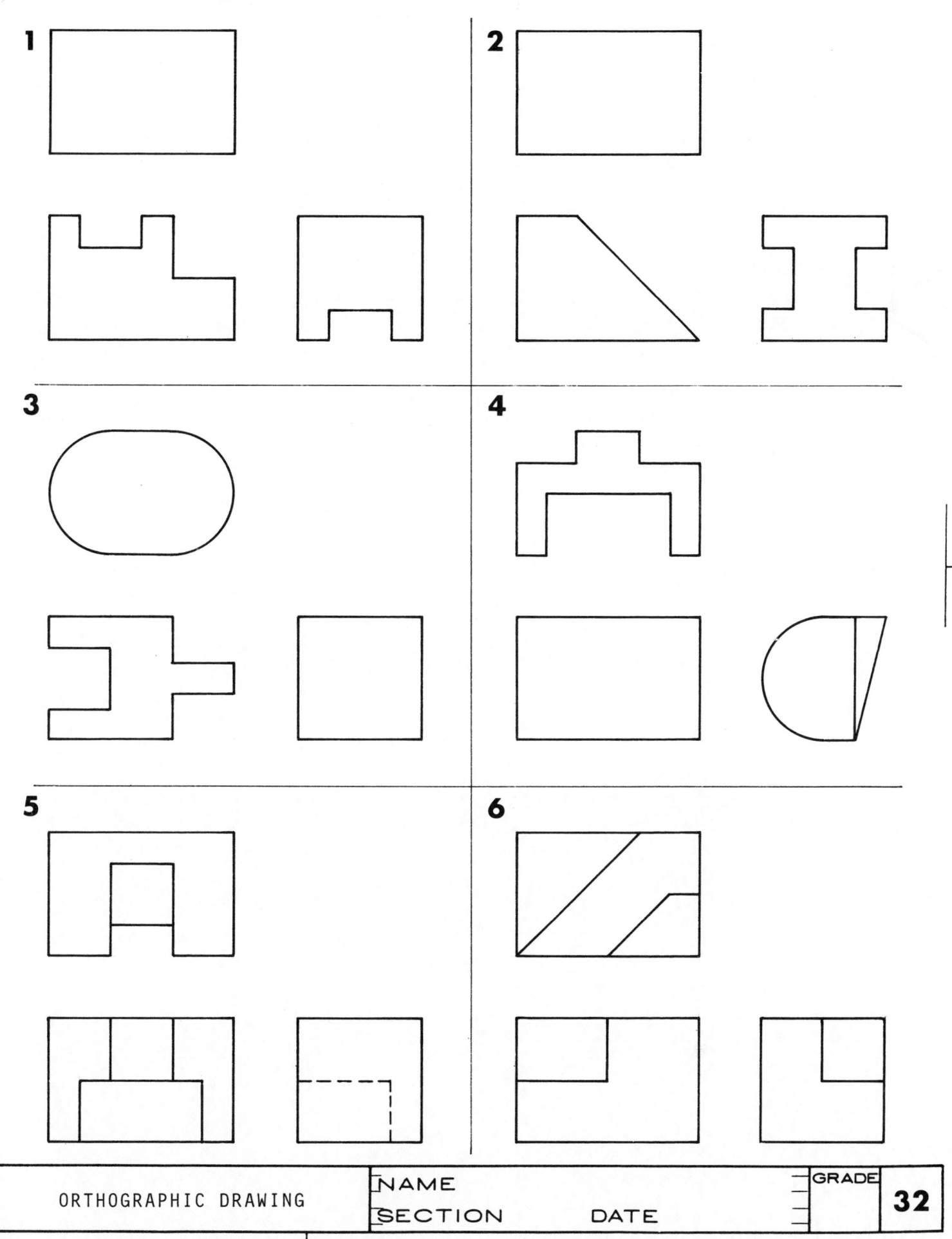

COMPLETE THE ORTHOGRAPHIC VIEWS BY ADDING ANY MISSING LINES. REFER TO THE ORTHOGRAPHIC VIEWS AND COMPLETE THE ISOMETRIC PICTORIAL SKETCHES.

| ISOMETRIC SKETCHING | NAME SECTION DATE | GRADE 33 |

REFER TO THE ORTHOGRAPHIC VIEWS AND COMPLETE
THE ISOMETRIC PICTORIAL SKETCH FOR EACH PROBLEM.

1

2

ISOMETRIC SKETCHING	NAME	GRADE
	SECTION DATE	**34**

REFER TO THE ORTHOGRAPHIC VIEWS AND COMPLETE
THE ISOMETRIC PICTORIAL SKETCH FOR EACH PROBLEM.

1

2

ISOMETRIC SKETCHING

35

COMPLETE THE ORTHOGRAPHIC VIEWS BY ADDING ANY MISSING LINES. REFER TO THE ORTHOGRAPHIC VIEWS AND COMPLETE THE ISOMETRIC PICTORIAL SKETCHES.

1

2

| ISOMETRIC SKETCHING | NAME | GRADE |
| | SECTION   DATE | 36 |

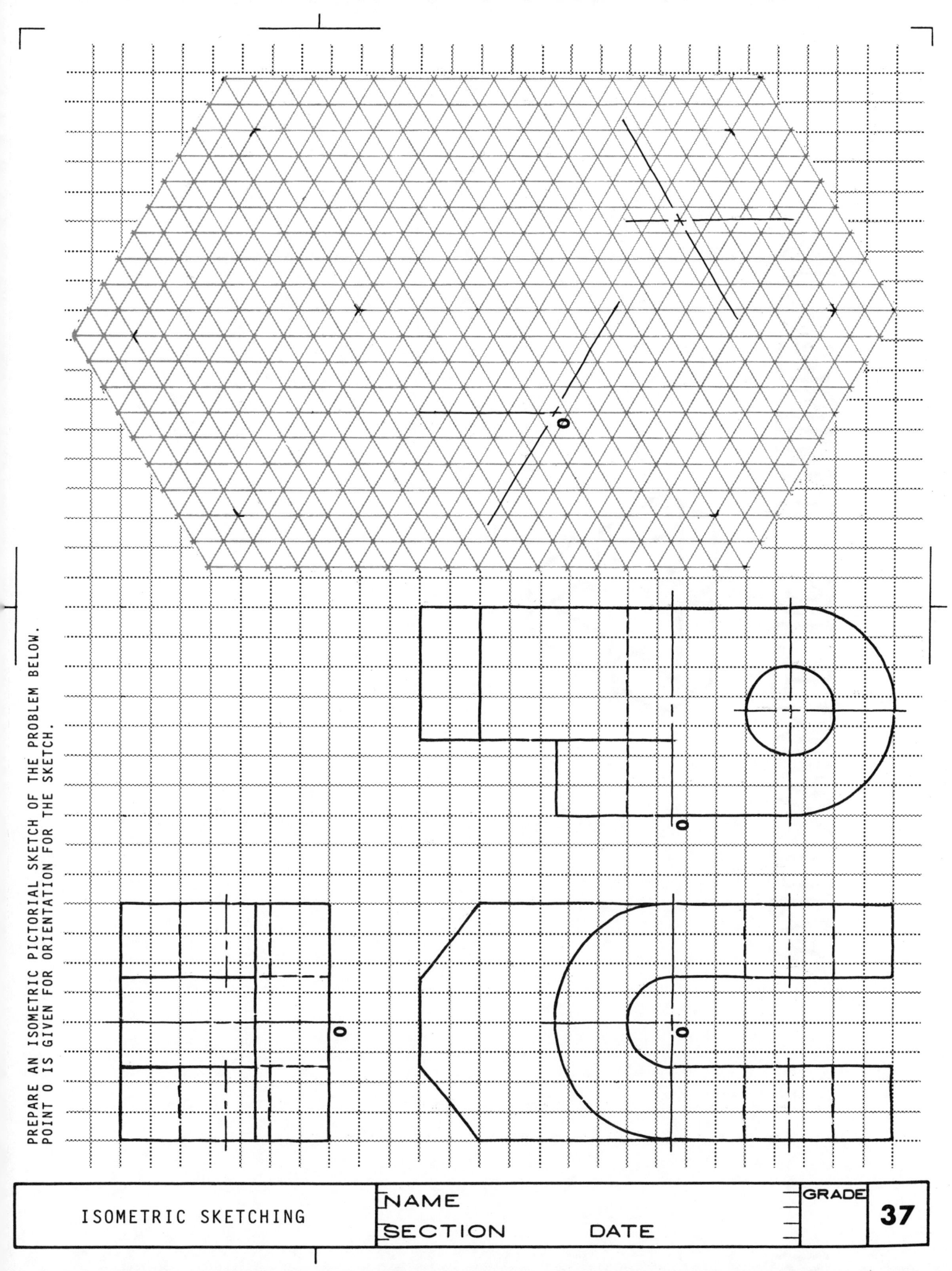

USING THE DIVIDERS TO TRANSFER MEASUREMENTS, CONSTRUCT
AN ISOMETRIC PICTORIAL DRAWING OF THE PROBLEM SHOWN BELOW.

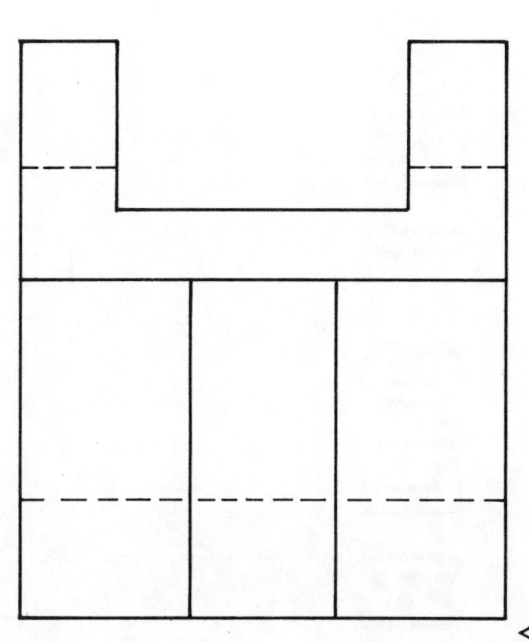
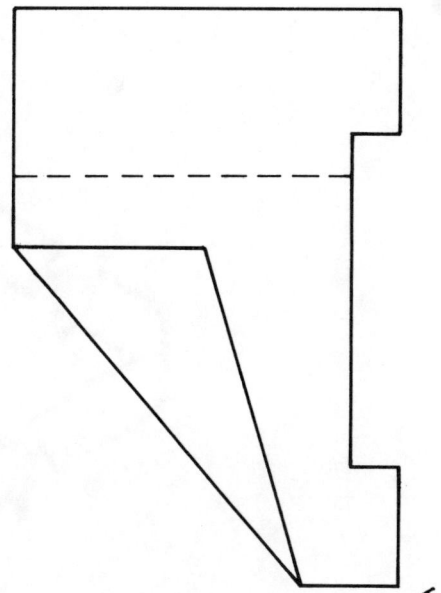

ISOMETRIC DRAWING	NAME	GRADE
	SECTION DATE	38

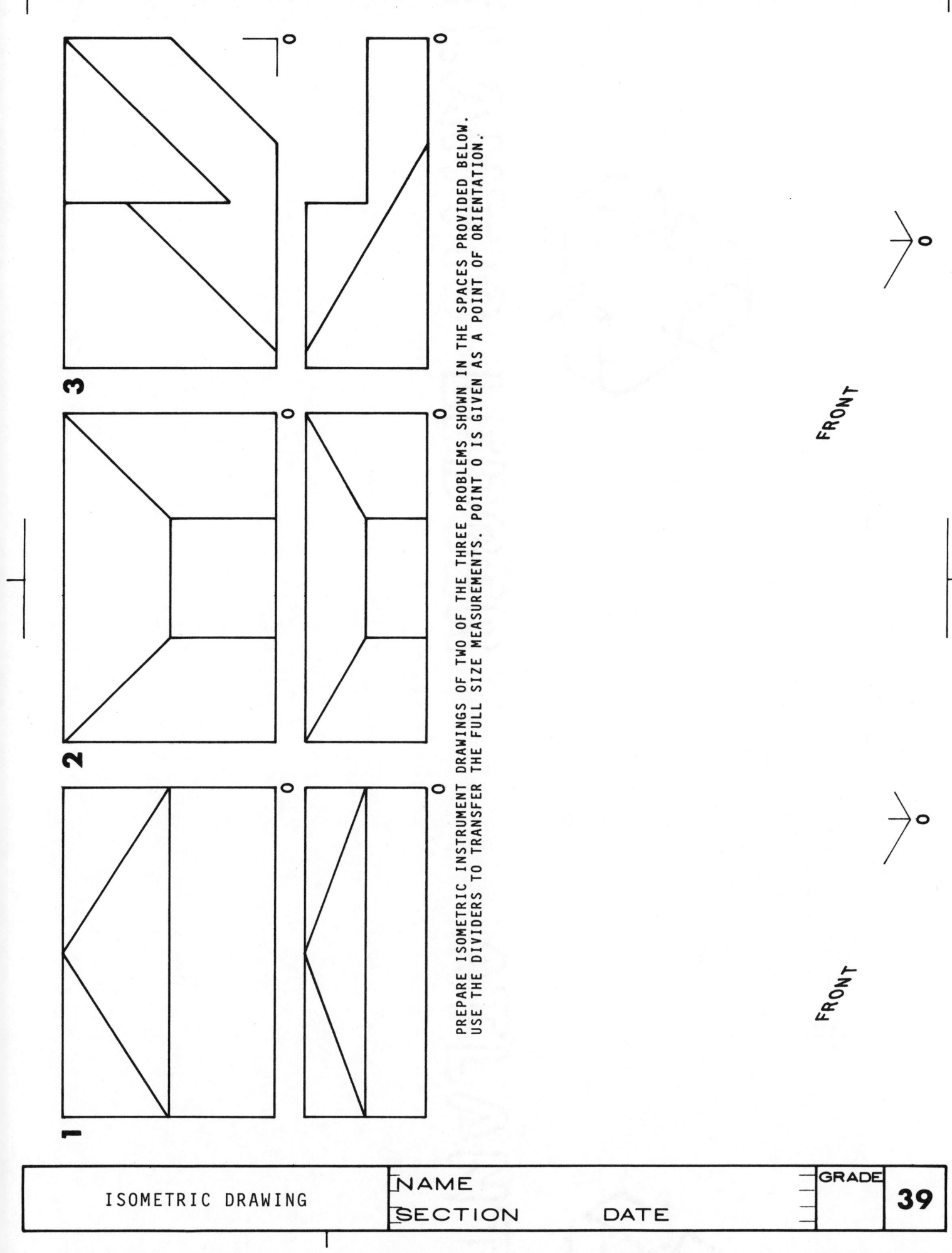

PREPARE A DOUBLE SIZE ISOMETRIC DRAWING OF THE PROBLEM AS SHOWN
IN THE SMALL DRAWING TO THE LEFT. USE THE DIVIDERS TO TRANSFER
THE MEASUREMENTS.

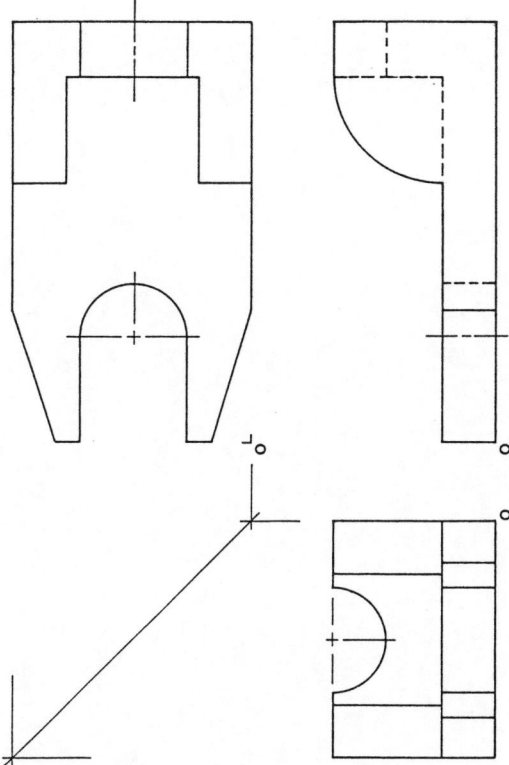

| ISOMETRIC DRAWING | NAME | GRADE |
| | SECTION   DATE | 40 |

ON A FRESH SHEET OF PAPER OR VELLUM FROM THE BACK OF THE WORKBOOK, PREPARE AN ISOMETRIC INSTRUMENT DRAWING OF THE PROBLEM ASSIGNED BY YOUR INSTRUCTOR.

1

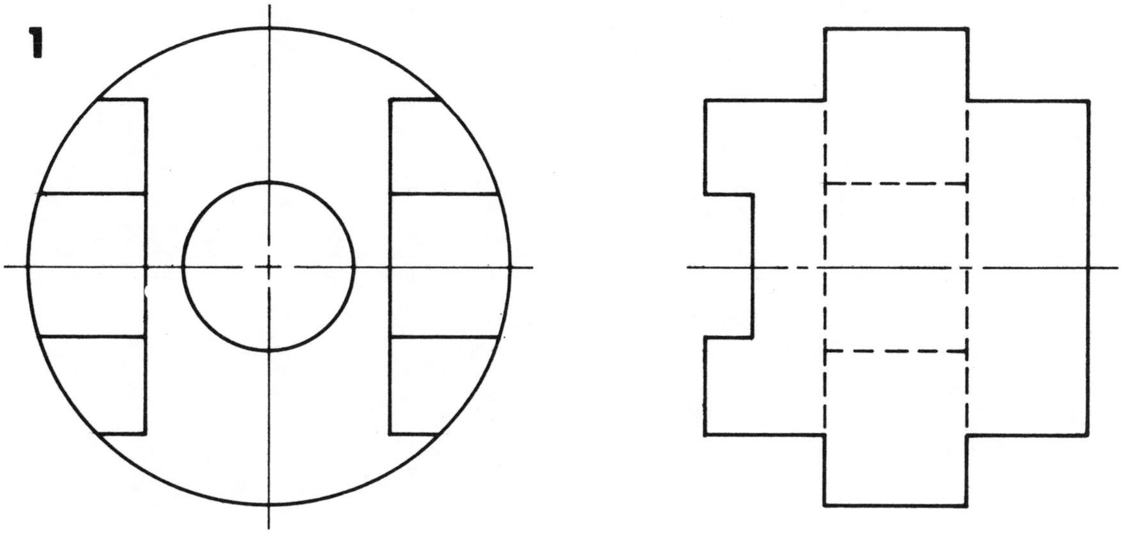

2

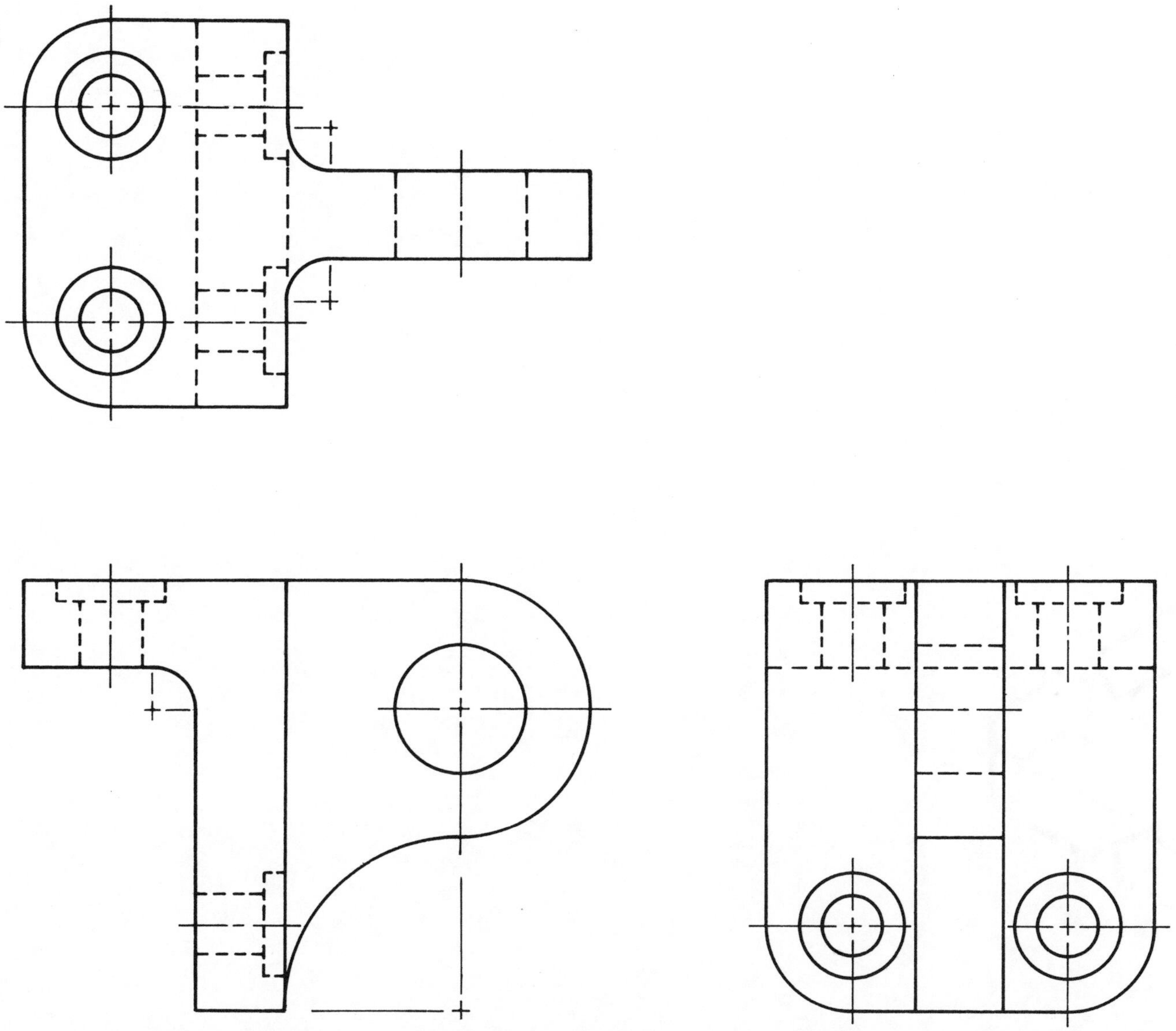

ISOMETRIC DRAWING

NAME
SECTION   DATE

GRADE

41

COMPLETE THE ORTHOGRAPHIC VIEWS BY ADDING ANY MISSING LINES. REFER TO
THE ORTHOGRAPHIC VIEWS AND COMPLETE THE OBLIQUE PICTORIAL SKETCHES.

1

2

OBLIQUE SKETCHING

NAME
SECTION   DATE

GRADE
42

REFER TO THE ORTHOGRAPHIC VIEWS AND COMPLETE
THE OBLIQUE PICTORIAL SKETCH FOR EACH PROBLEM.

1

2

| OBLIQUE SKETCHING | NAME | GRADE |
| | SECTION DATE | **43** |

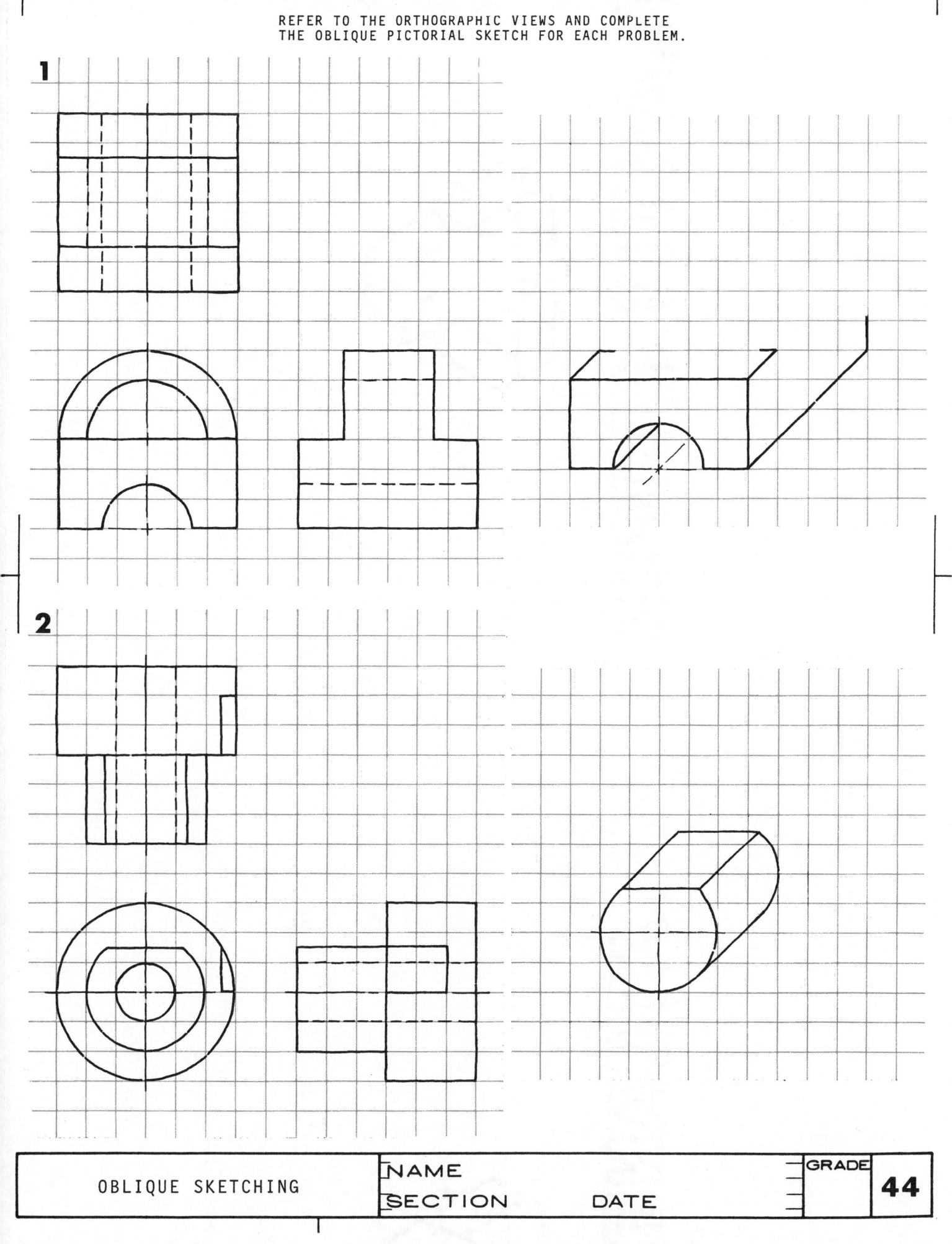

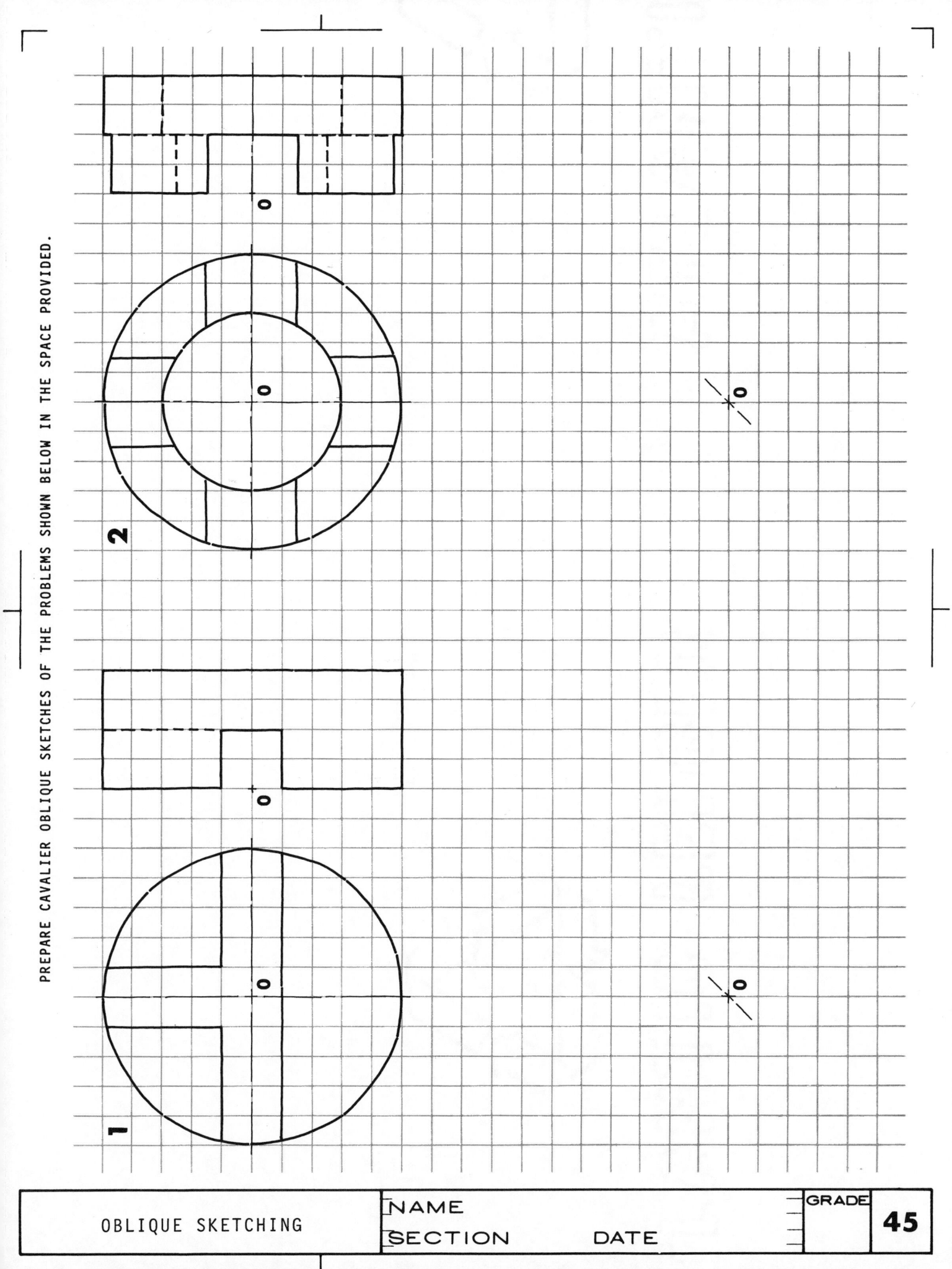

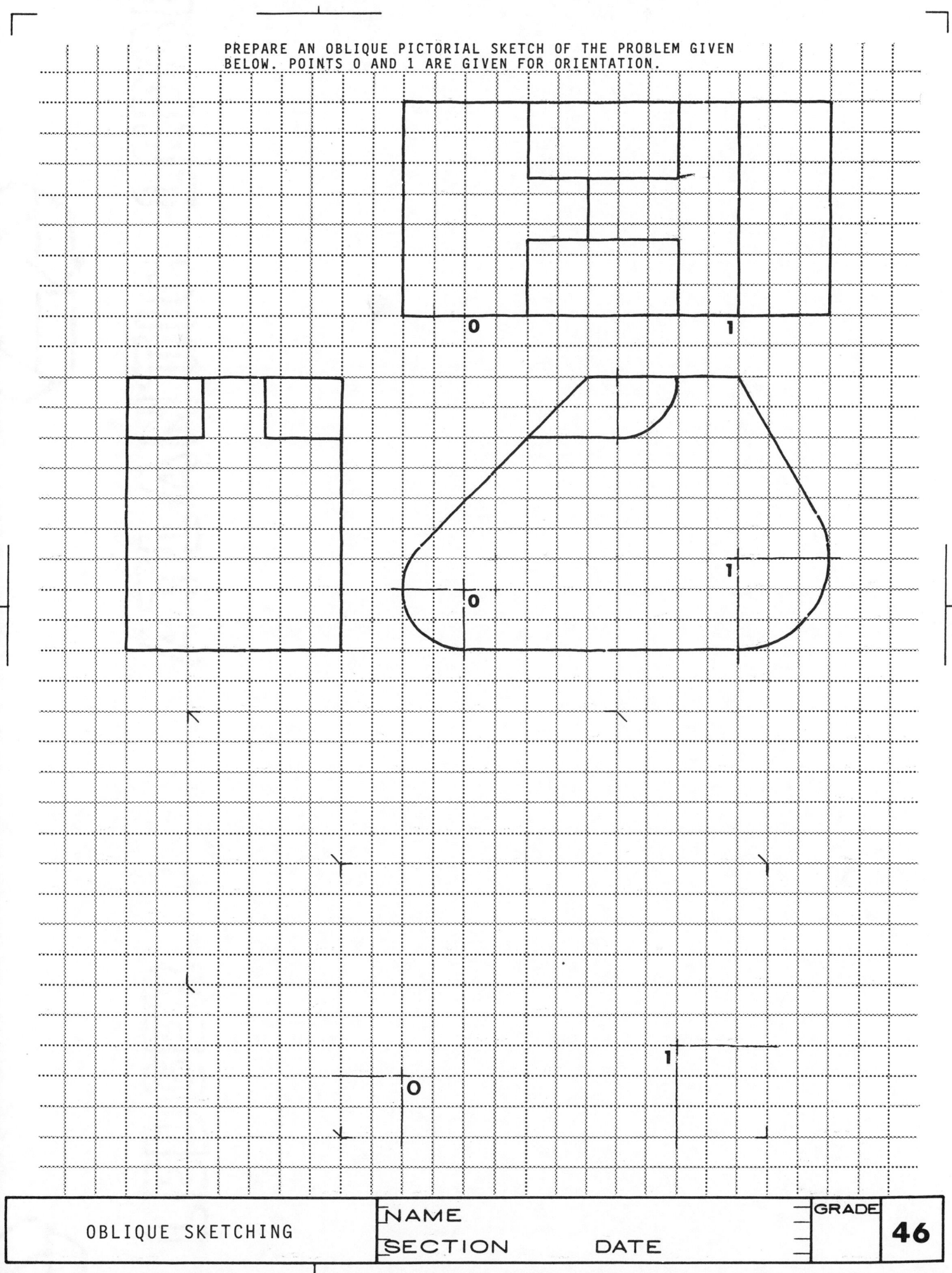

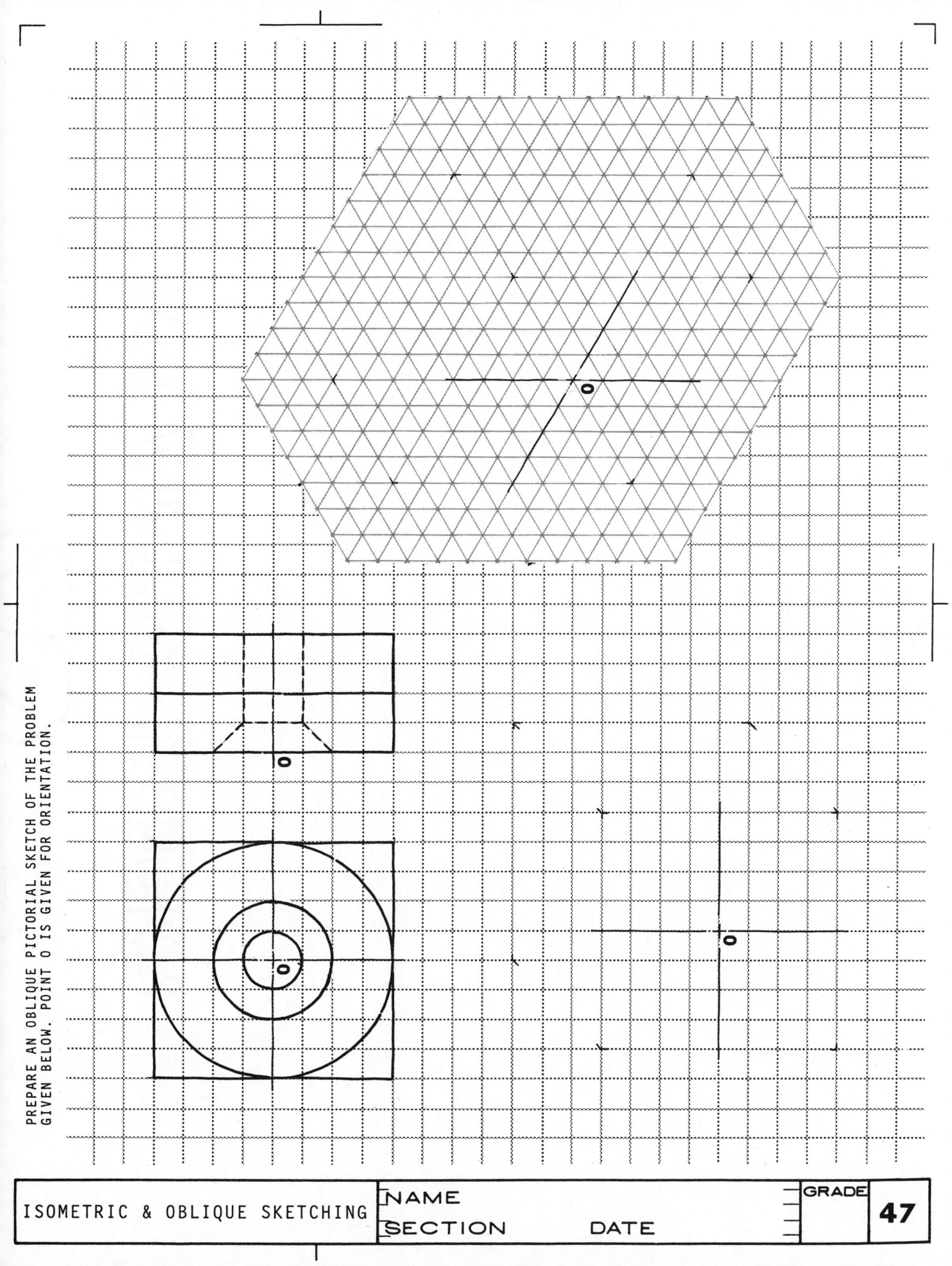

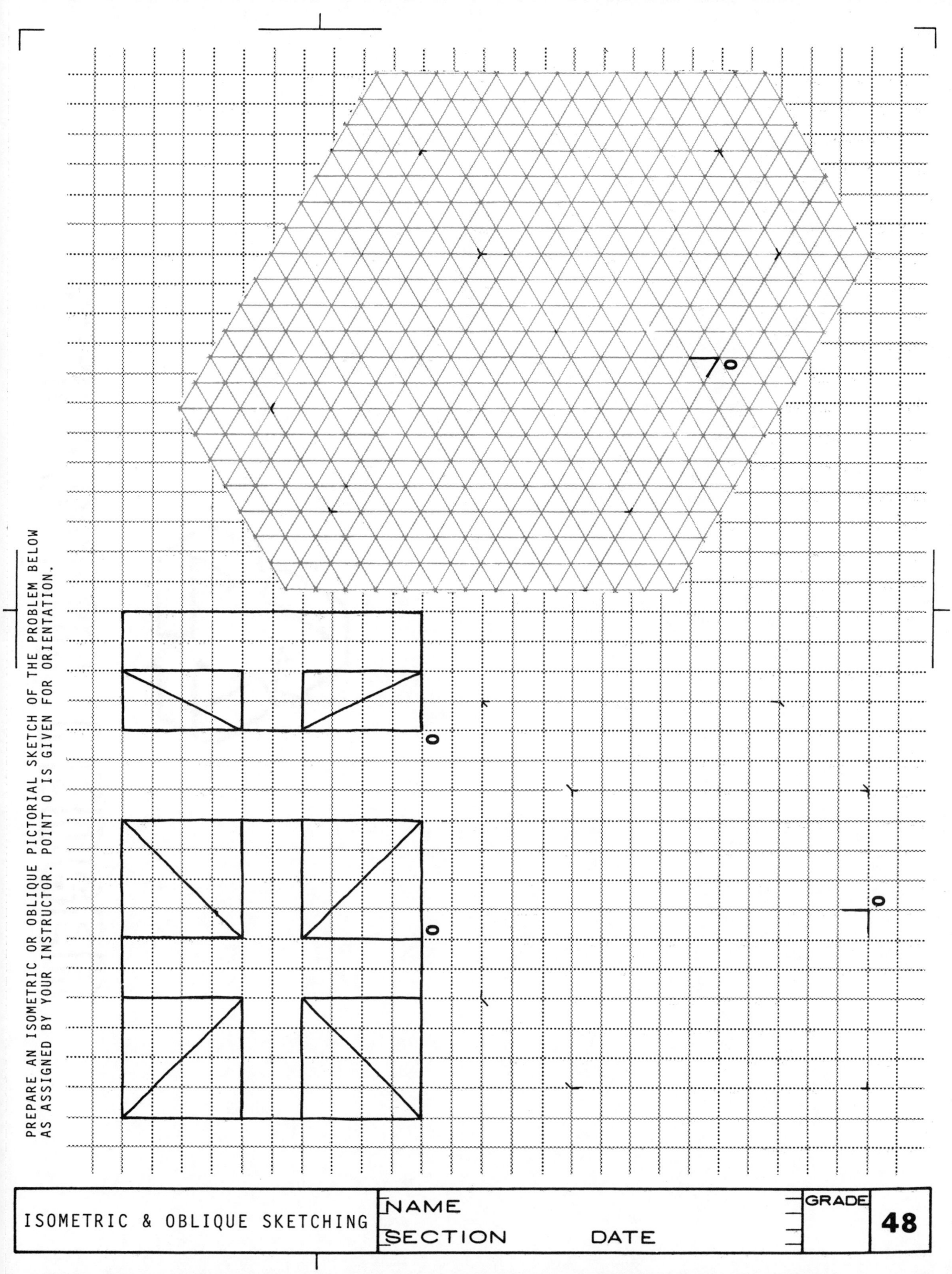

PREPARE A FULL-SIZE ISOMETRIC OR CAVALIER OBLIQUE DRAWING OF THE PROBLEM SHOWN BELOW. BEGIN THE DRAWING AT POINT O.

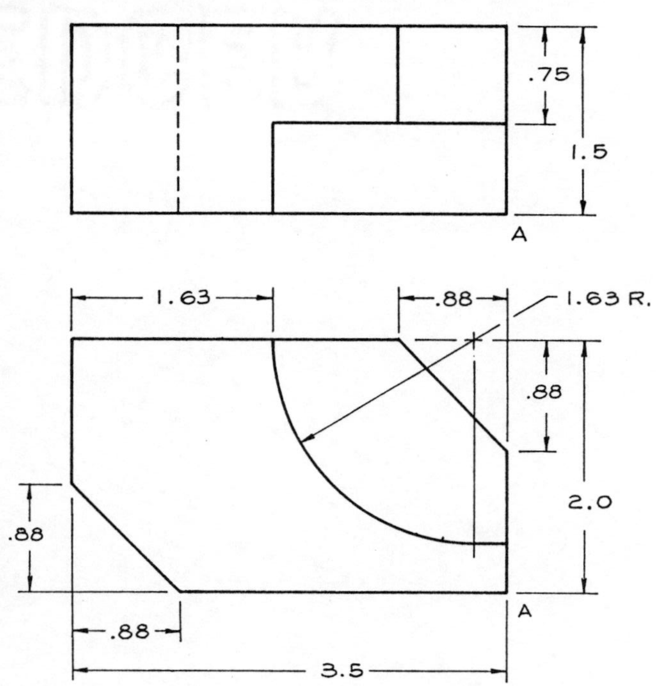

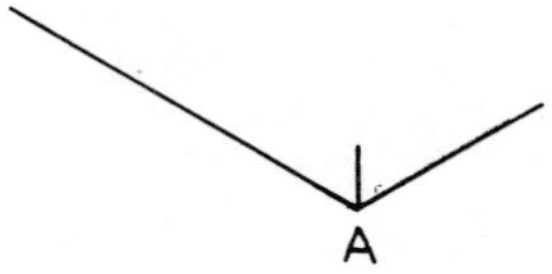

ISOMETRIC

OBLIQUE

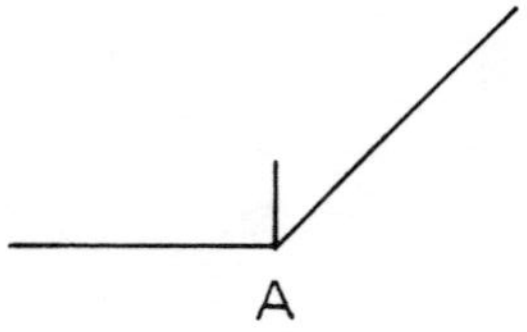

ISOMETRIC & OBLIQUE DRAWING	NAME	GRADE
	SECTION DATE	49

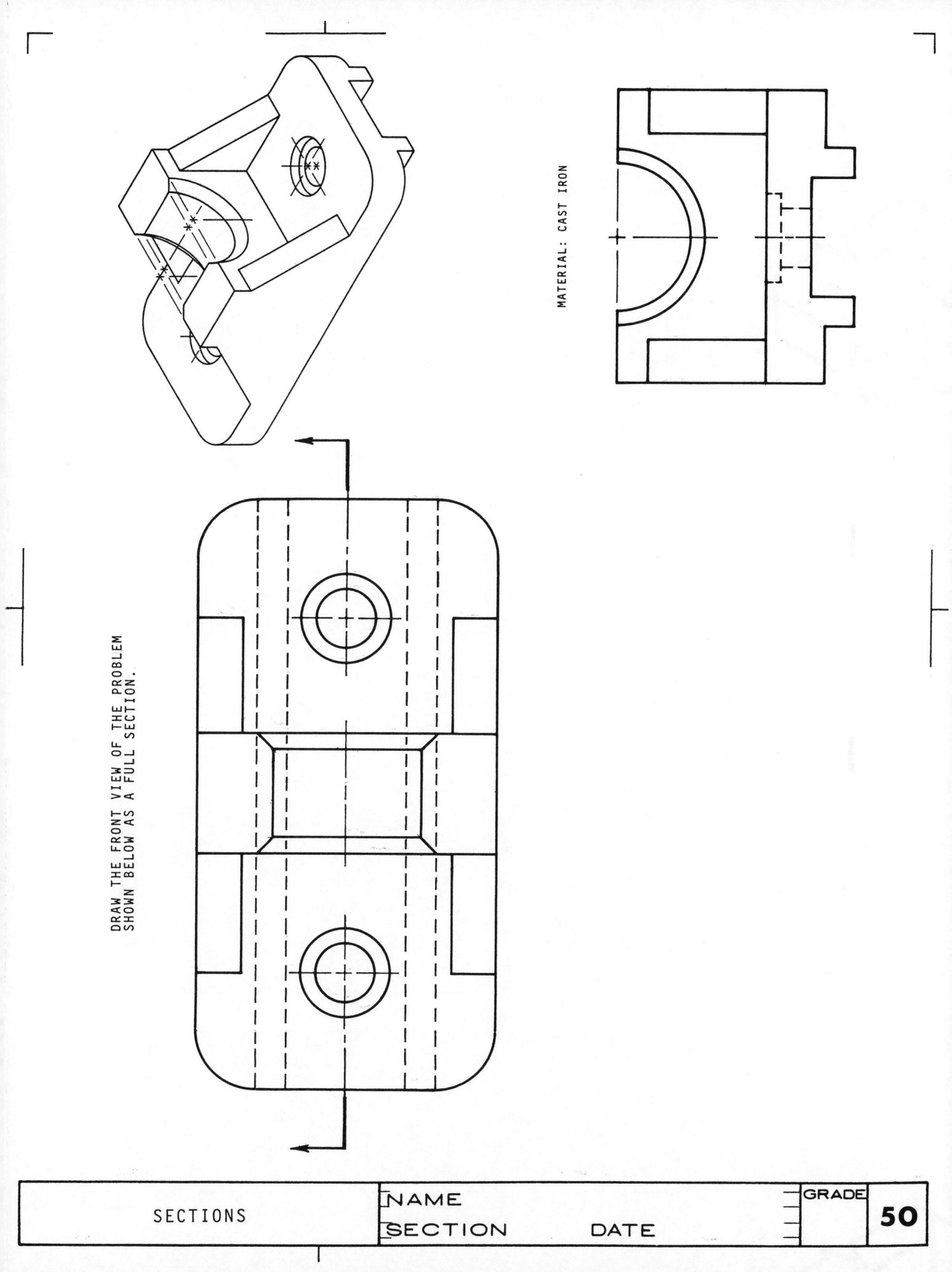

MATERIAL: MAGNESIUM

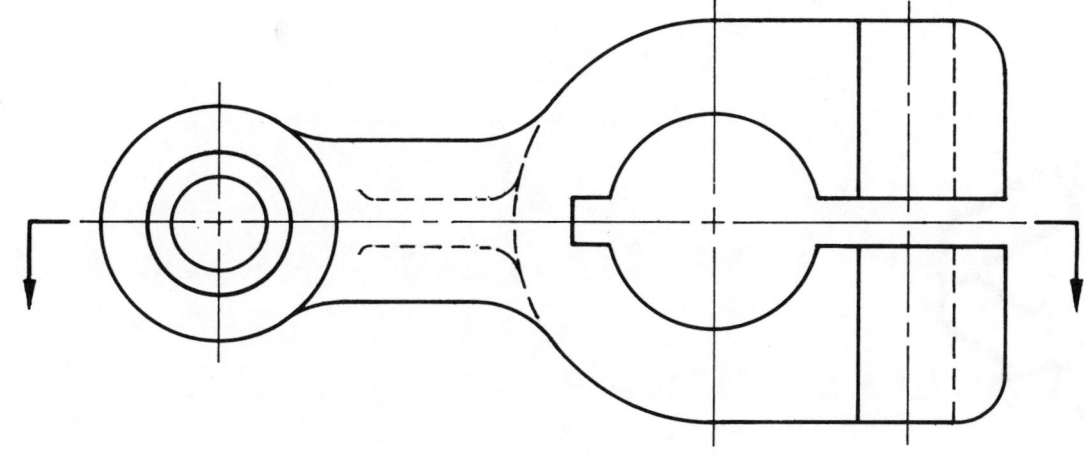

DRAW THE RIGHT SIDE VIEW OF THE
PROBLEM SHOWN AS A FULL SECTION.

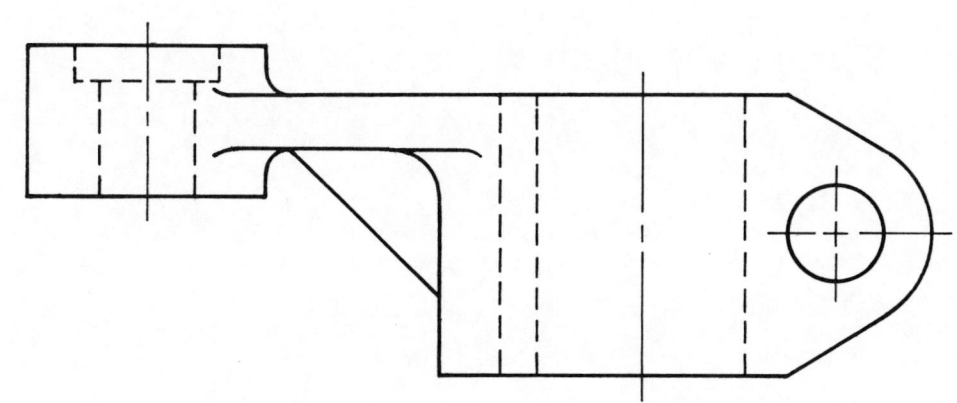

SECTIONS	NAME	GRADE	51
	SECTION DATE		

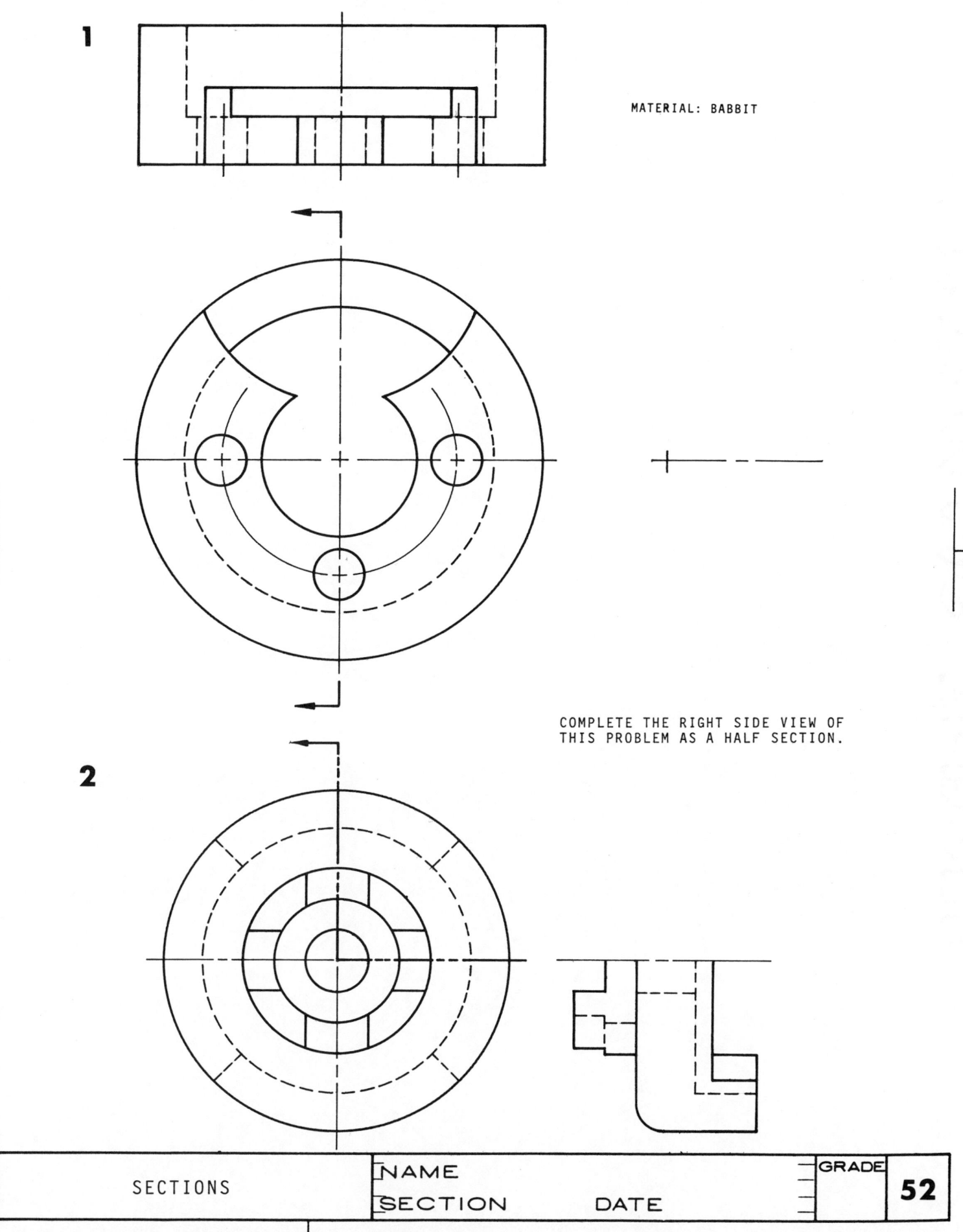

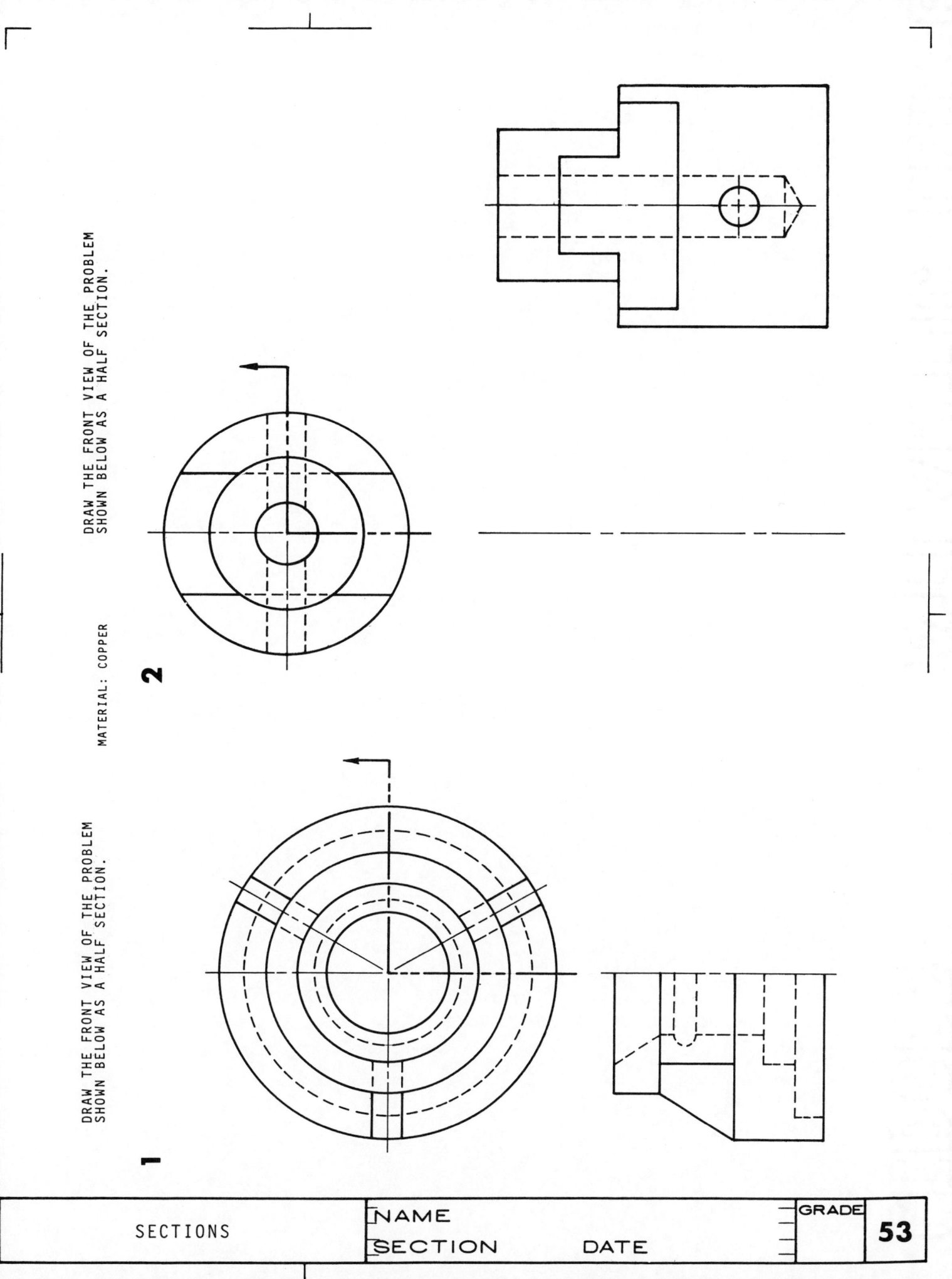

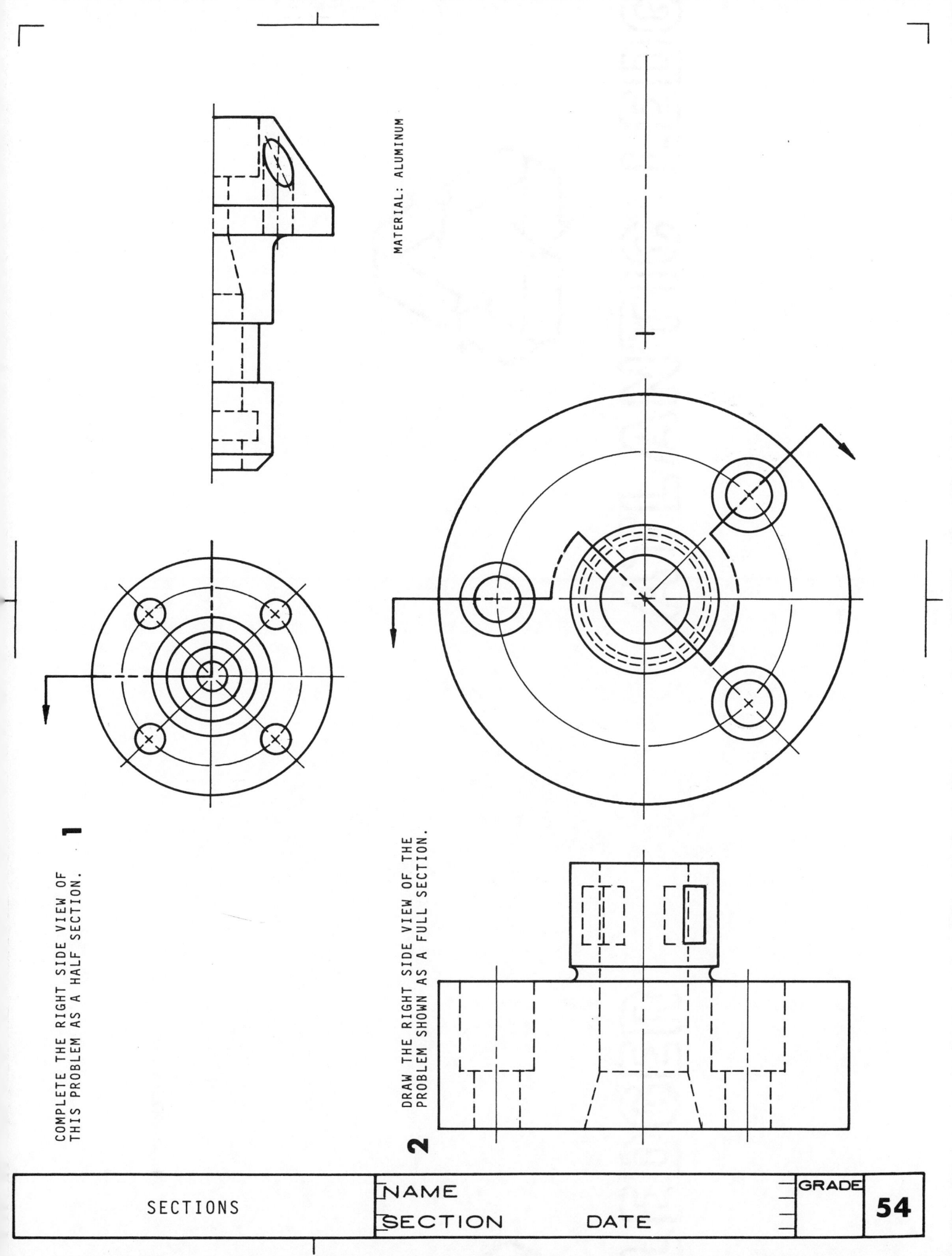

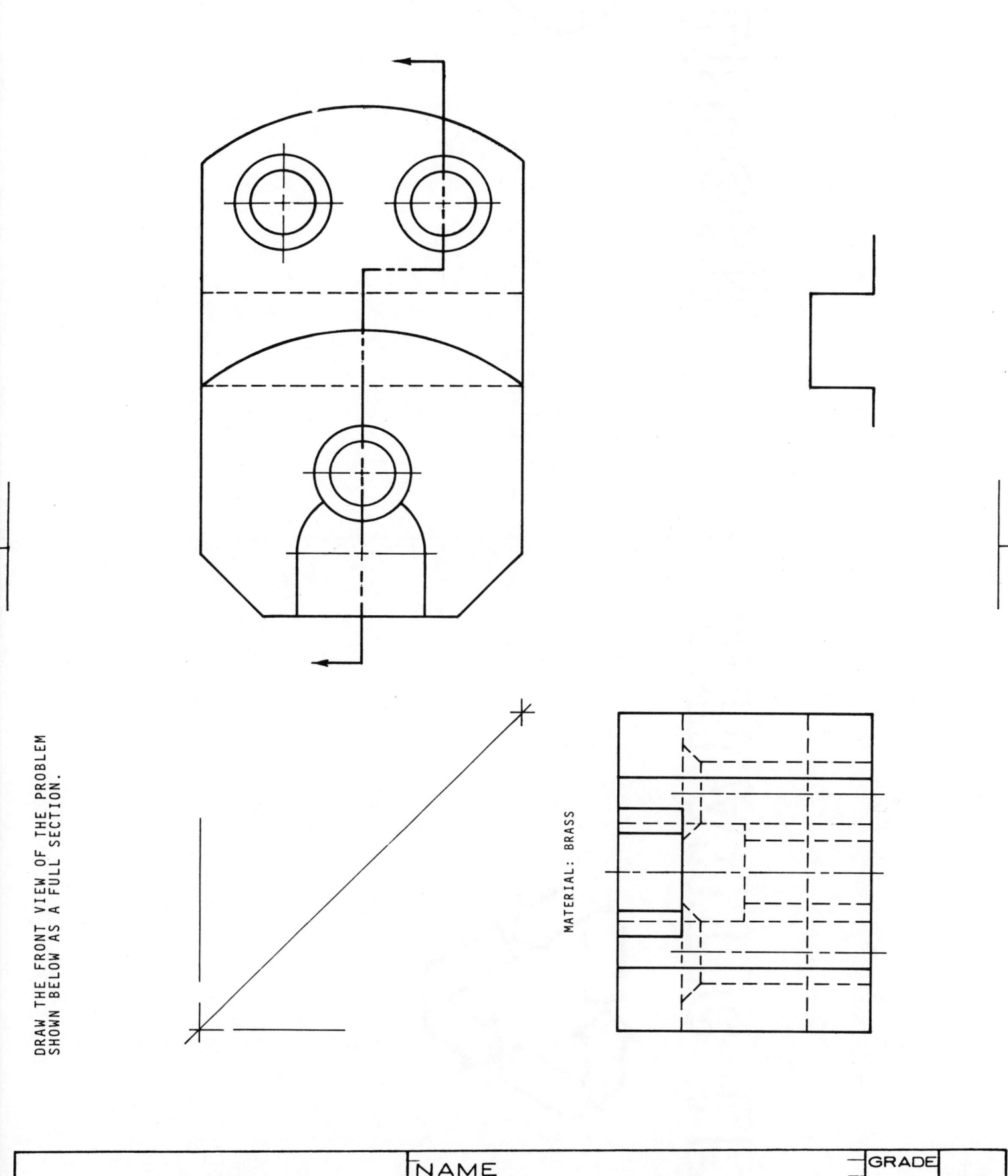

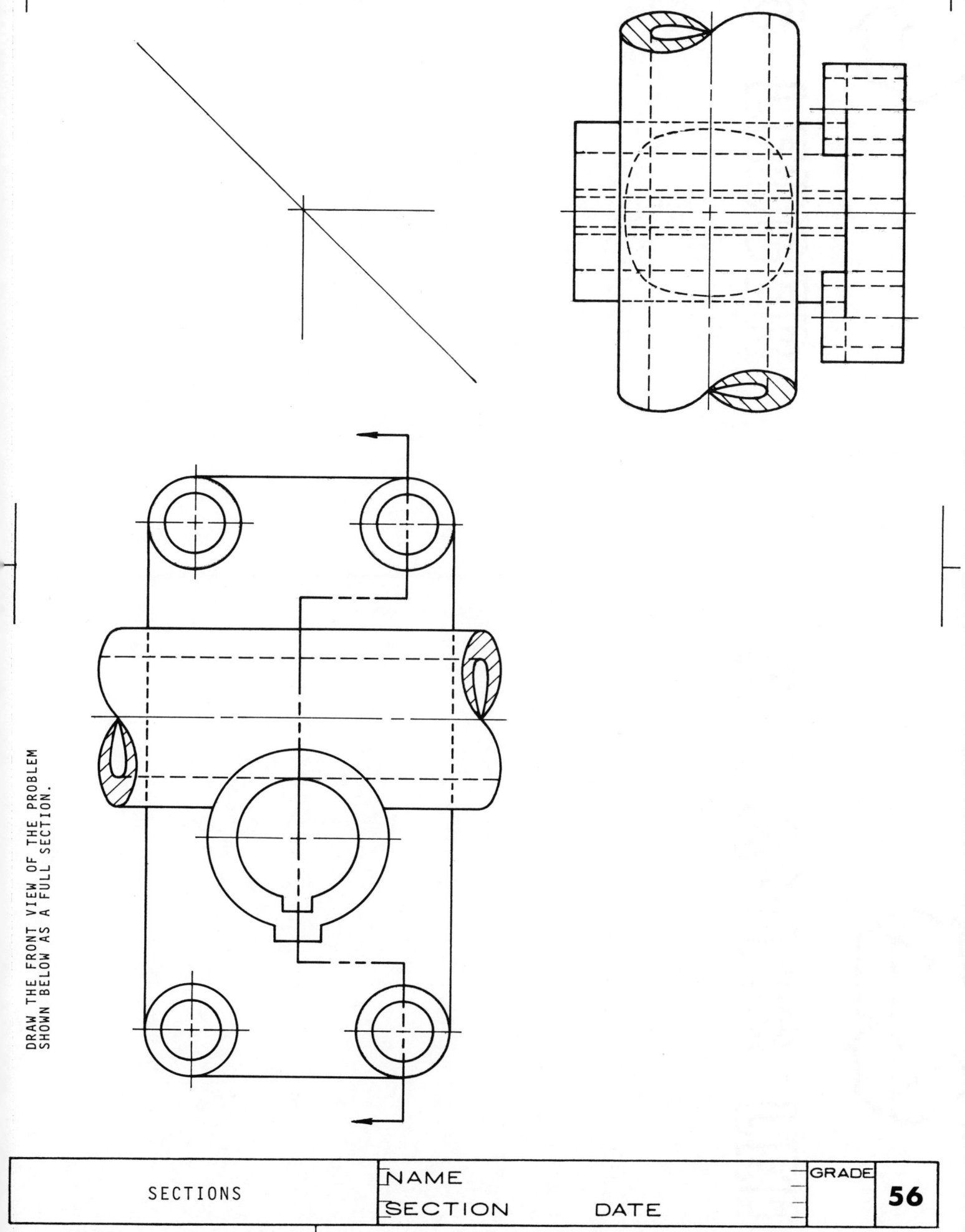

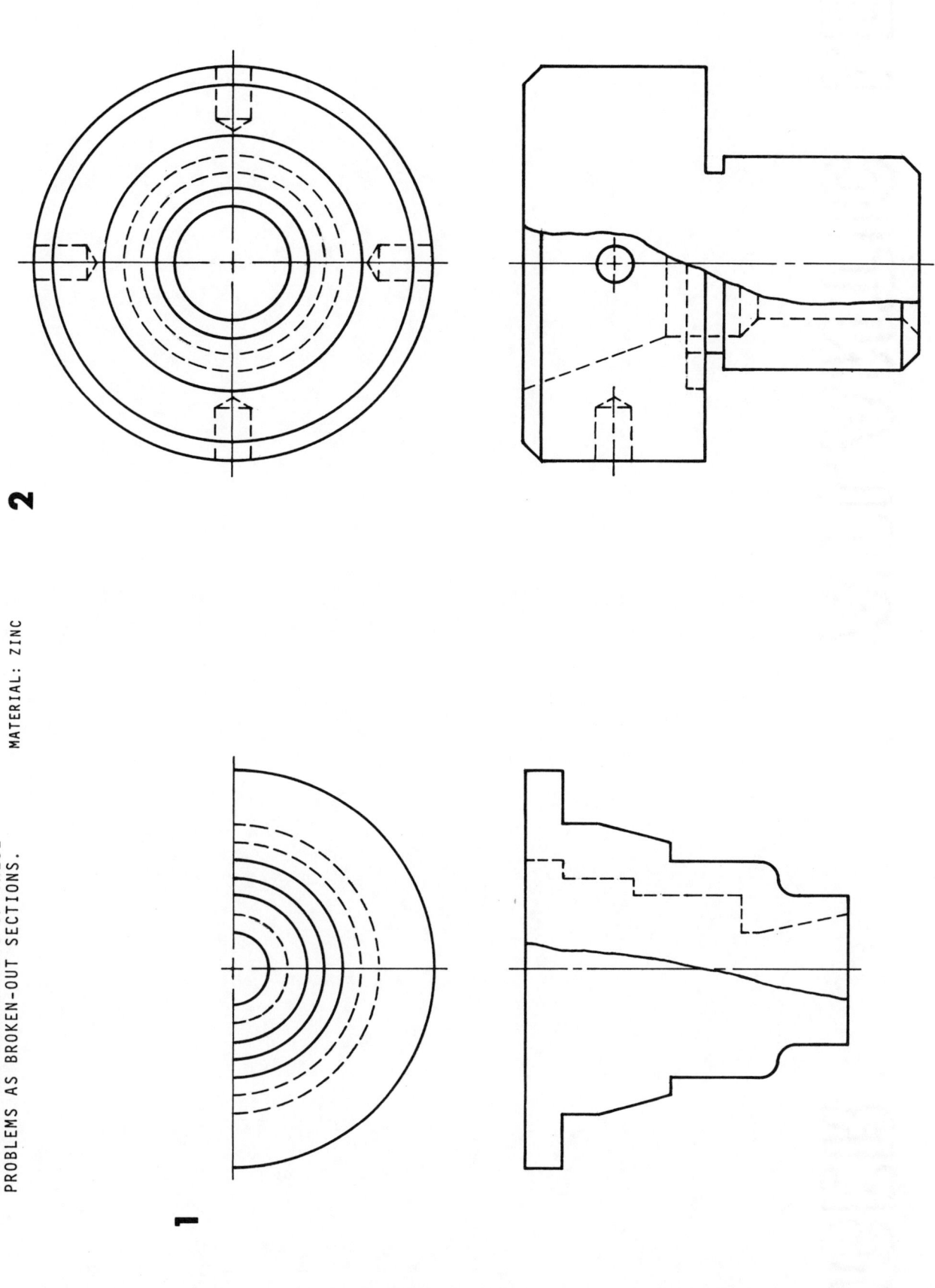

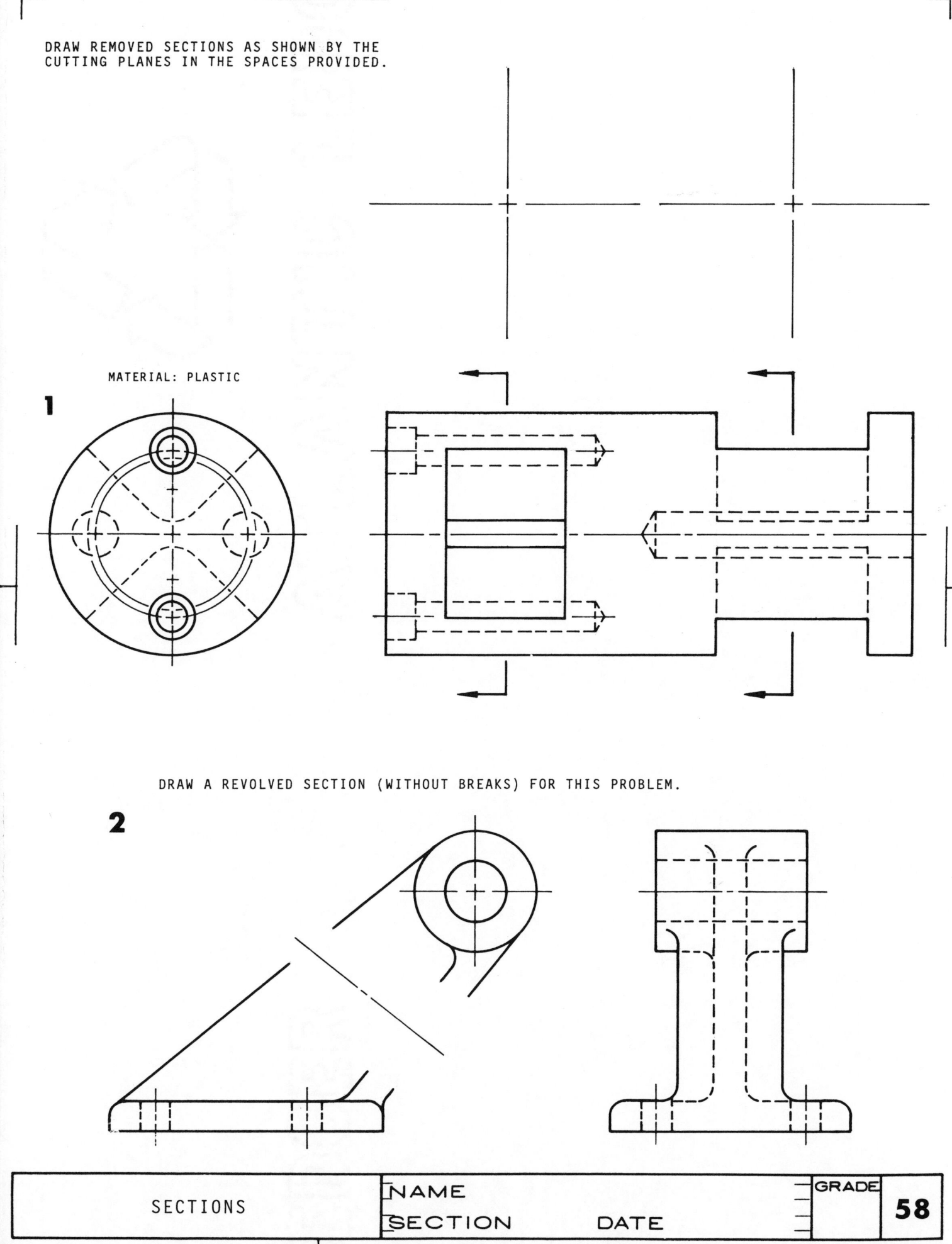

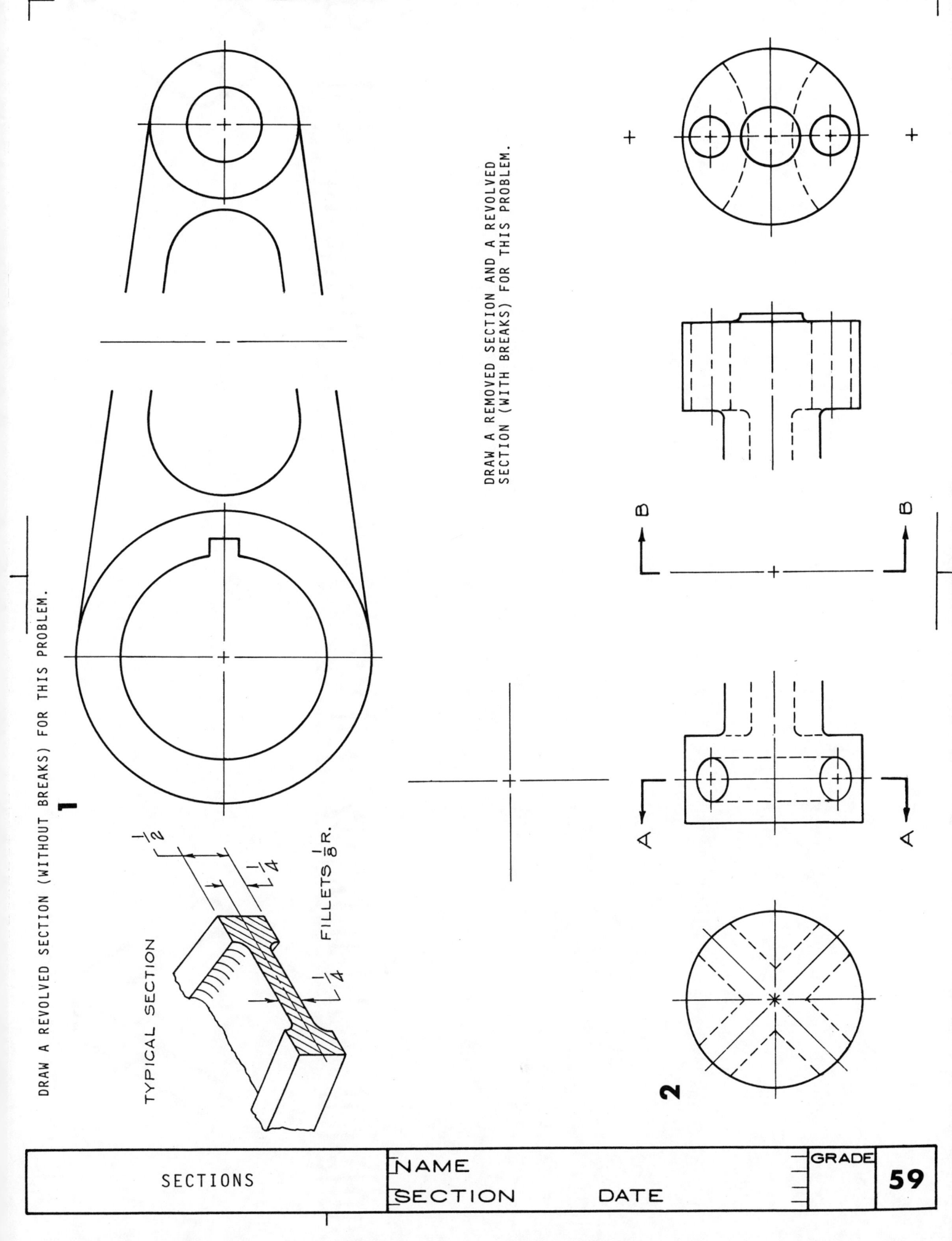

DRAW THE COMPLETE FRONT AND TOP VIEWS OF THE PROBLEM AS SHOWN TO THE RIGHT IN THE SPACES GIVEN BELOW.   DRAWING SCALE: 1" = 1"

NOTE: DRAW THE FRONT VIEW AS A SECTIONED VIEW TO CONFORM TO THE CUTTING PLANE A-A AS GIVEN.

MATERIAL: STEEL

TOP VIEW

FRONT VIEW

SECTIONS

NAME

SECTION   DATE

GRADE

60

DIMENSION THE PROBLEMS BELOW USING THE DECIMAL DIMENSIONING SYSTEM.
USE METRIC OR INCH VALUES. EACH GRID UNIT REPRESENTS 6mm OR .25 INCH.

1

2

3

DIMENSIONING	NAME		GRADE	61
	SECTION	DATE		

DIMENSION THE PROBLEMS BELOW USING THE DECIMAL DIMENSIONING SYSTEM.
USE METRIC OR INCH VALUES. EACH GRID UNIT REPRESENTS 6mm OR .25 INCH.

1

2

DIMENSIONING	NAME	GRADE
	SECTION DATE	**62**

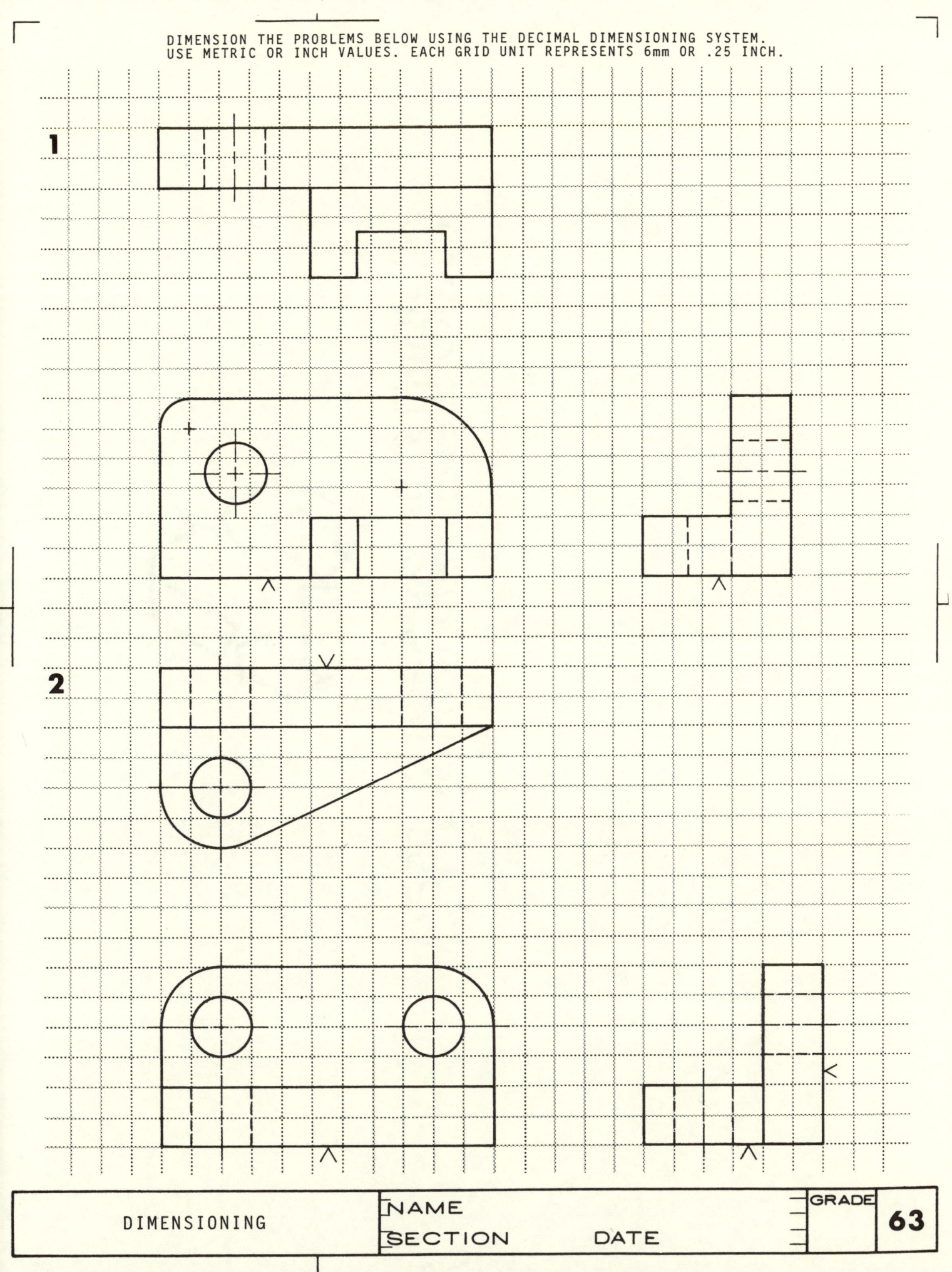

DIMENSION THE DRAWINGS BELOW USING METRIC OR DECIMAL-INCH MEASUREMENTS.
NOTE: SCALE THE DRAWINGS TO OBTAIN MEASUREMENTS.

1

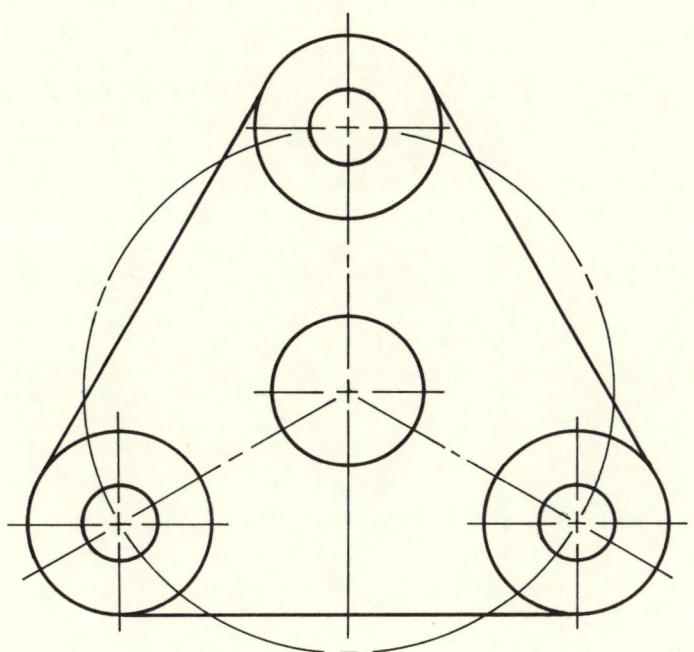
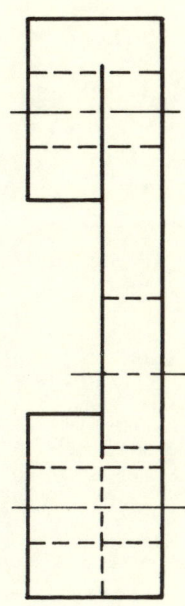

USE ANGULAR LOCATION DIMENSIONS FOR PROBLEM 1.
USE COORDINATE LOCATION DIMENSIONS FOR PROBLEM 2.

2

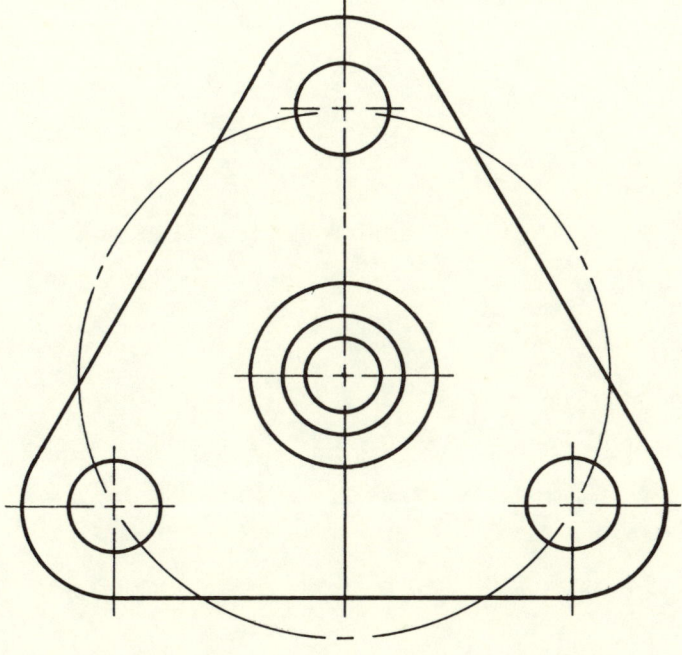
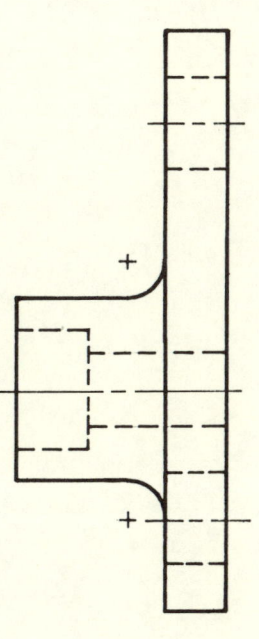

DIMENSIONING | NAME | GRADE
SECTION   DATE | 64

DIMENSION THE DRAWINGS BELOW USING METRIC OR DECIMAL-INCH MEASUREMENTS.

1

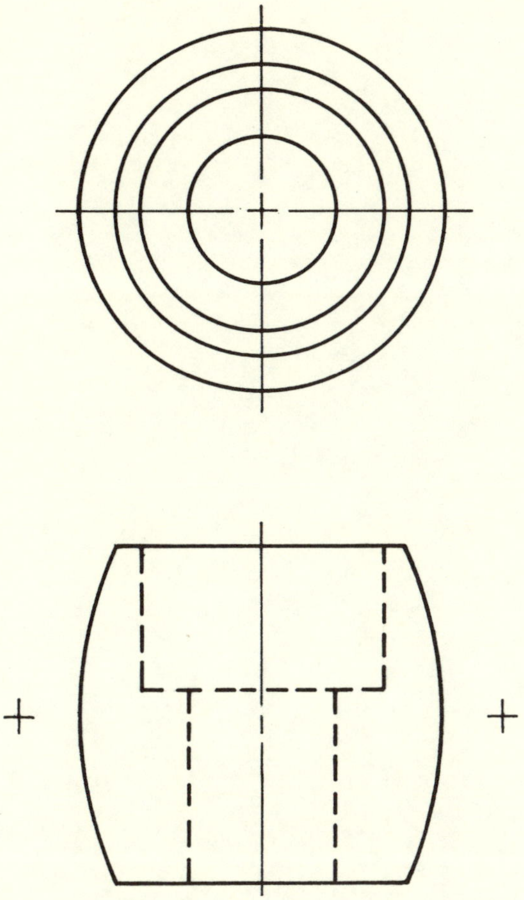

2

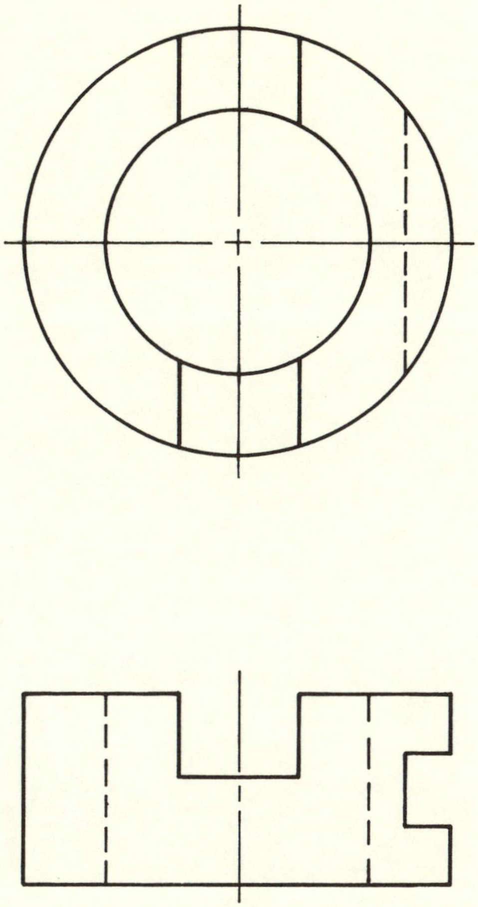

NOTE: SCALE THE DRAWINGS TO OBTAIN MEASUREMENTS.

3

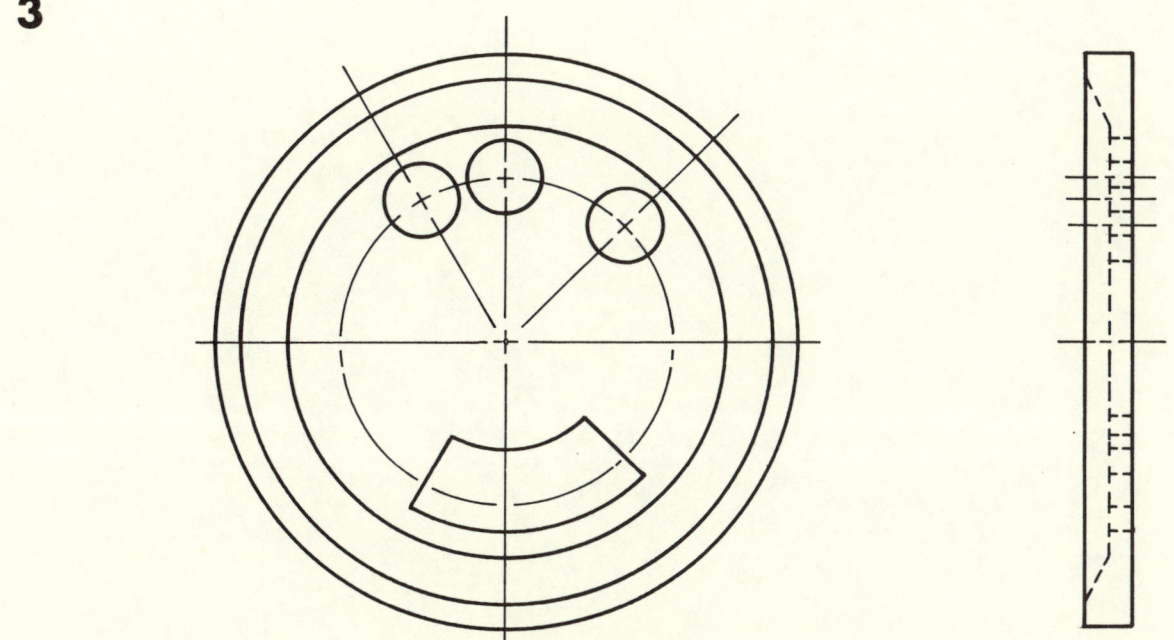

DIMENSIONING

NAME
SECTION   DATE

GRADE
65

DIMENSION THE DRAWINGS BELOW USING METRIC OR DECIMAL-INCH MEASUREMENTS.

NOTE: SCALE THE DRAWINGS TO OBTAIN MEASUREMENTS.

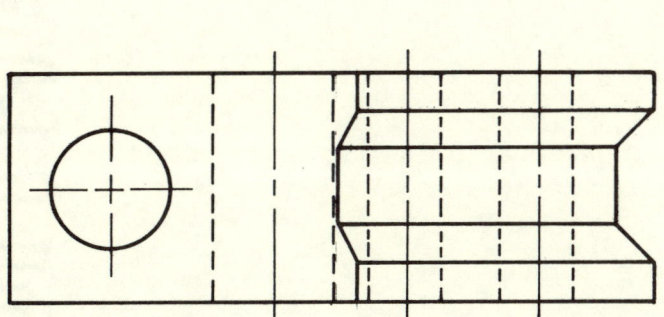
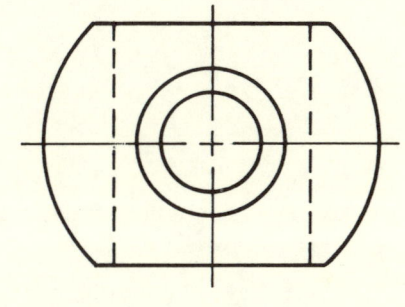
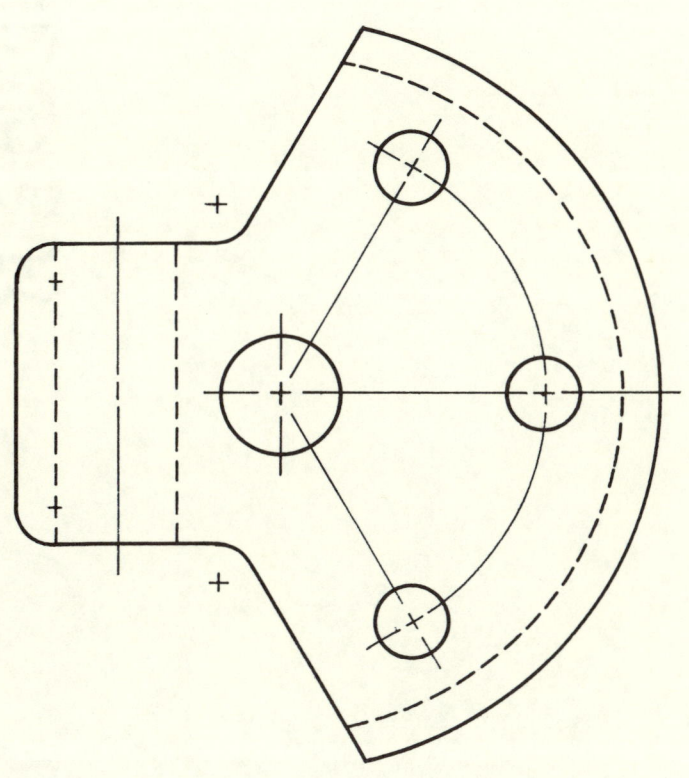
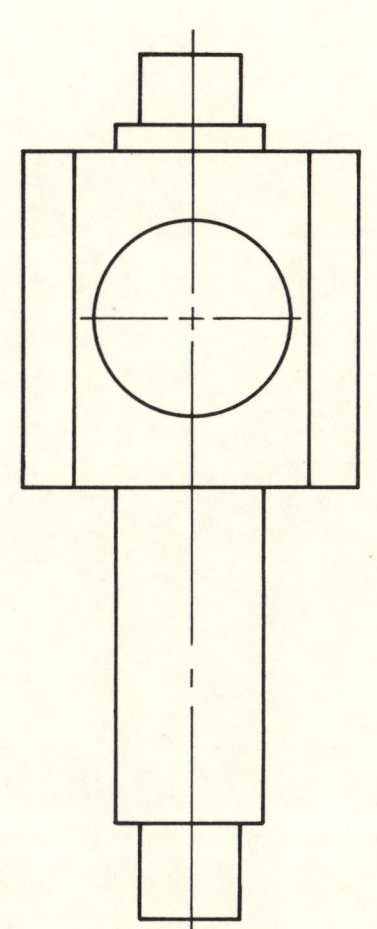

1

2

DIMENSIONING	NAME	GRADE	66
	SECTION DATE		

DETERMINE THE POINTS OF TANGENCY FOR THE PROBLEMS BELOW AND
FULLY DIMENSION USING METRIC OR DECIMAL-INCH MEASUREMENTS.

1

2

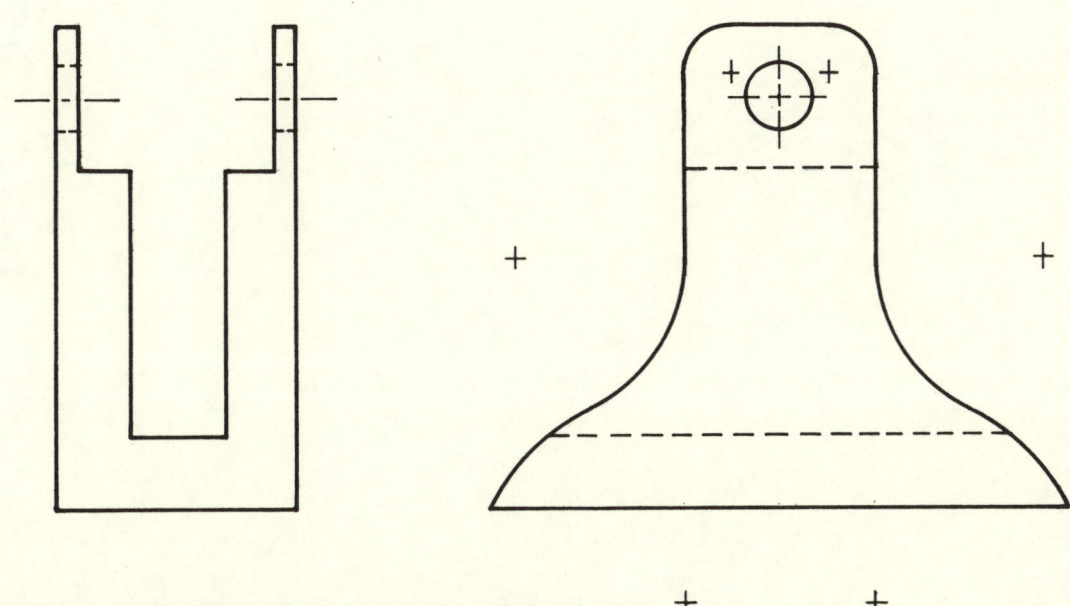

NOTE: SCALE THE DRAWINGS TO OBTAIN MEASUREMENTS.

DIMENSIONING	NAME	GRADE	67
	SECTION DATE		

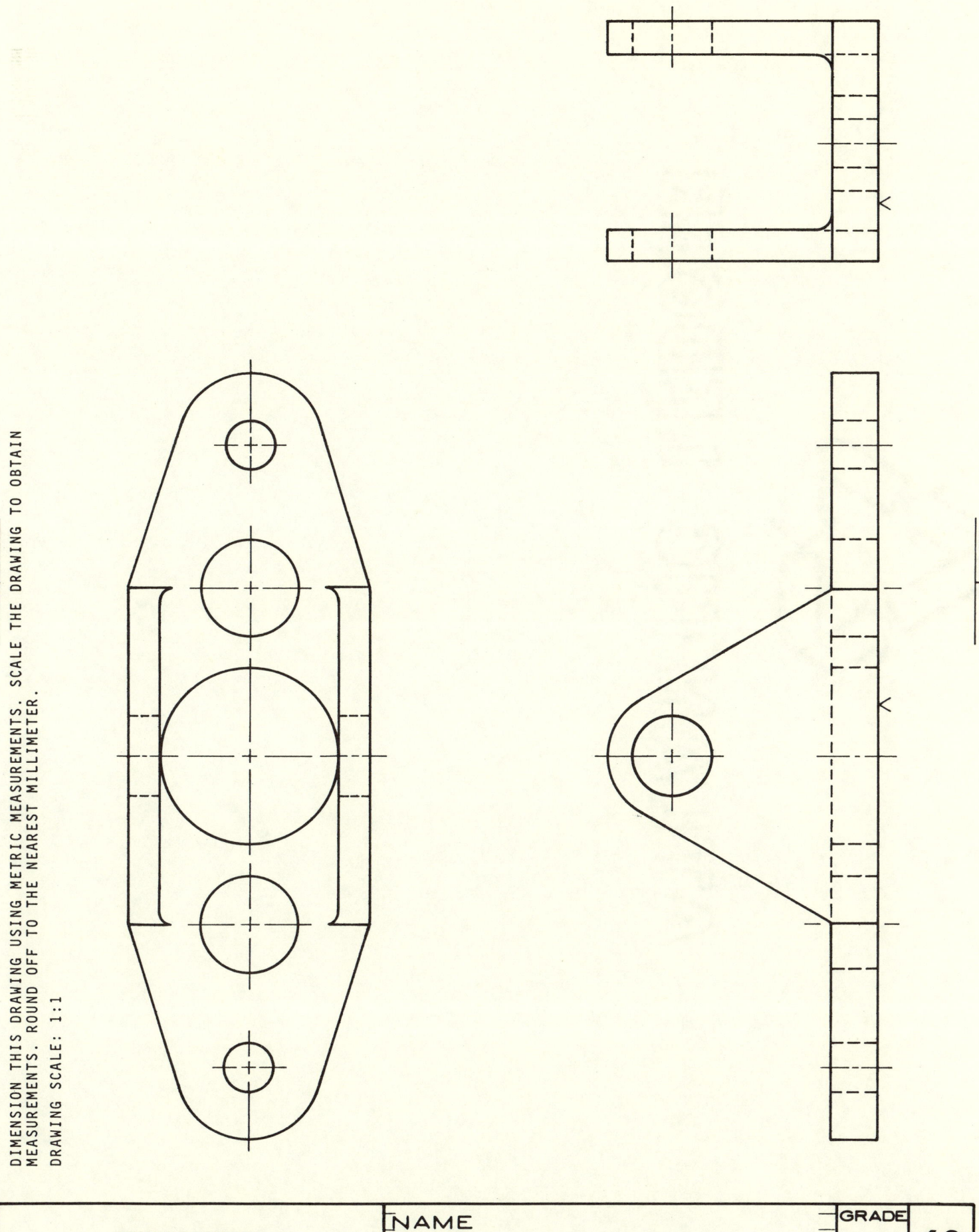

REFER TO THE TABLE OF FITS LOCATED IN THE TEXTBOOK APPENDIX AND
CALCULATE THE LIMIT DIMENSIONS FOR THE ASSEMBLIES SHOWN BELOW.
RECORD THE DIMENSIONS IN THE GUIDELINES PROVIDED.

BASIC DIA. 0.500 CLASS RC2 FIT

BASIC DIA. 0.750 CLASS LT3 FIT

BASIC DIA. 0.875 CLASS FN1 FIT

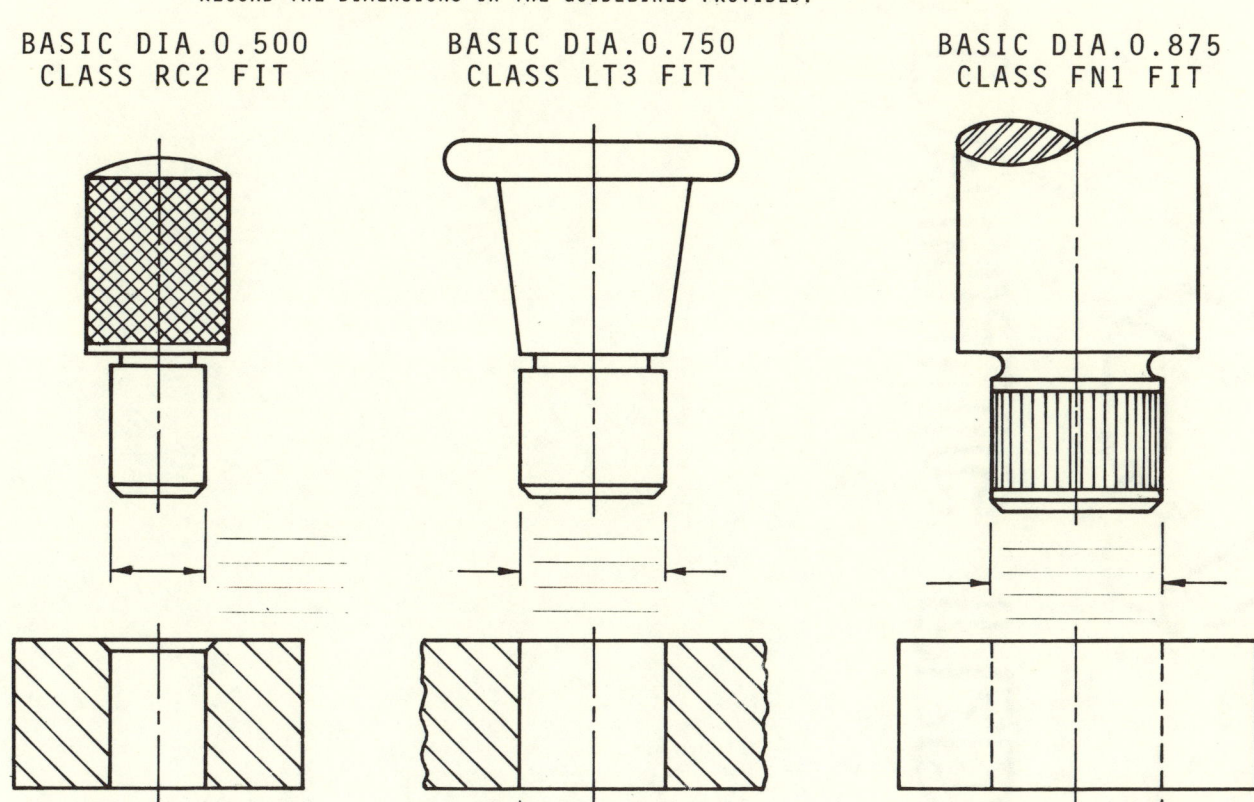

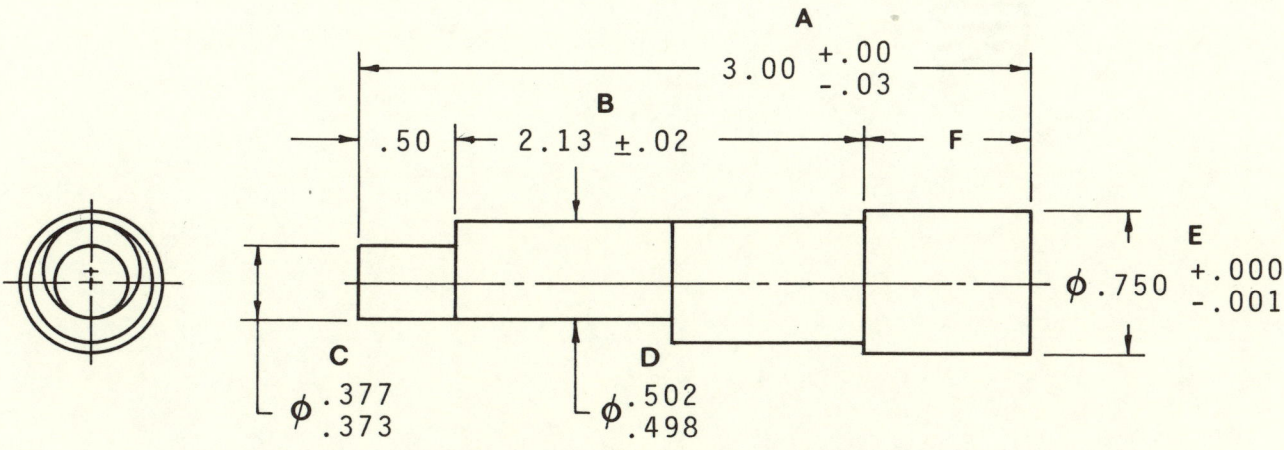

COMPLETE THE TABLE FROM THE INFORMATION GIVEN ABOVE

		A	B	C	D	E	F
BASIC SIZE							XXXX
TOLERANCE							XXXX
LIMITS OF SIZE	MAX						
	MIN						

TOLERANCES

REFER TO THE TABLE OF FITS LOCATED IN THE TEXTBOOK APPENDIX AND
CALCULATE AND RECORD THE MISSING DIMENSIONS FOR THE CHART BELOW.

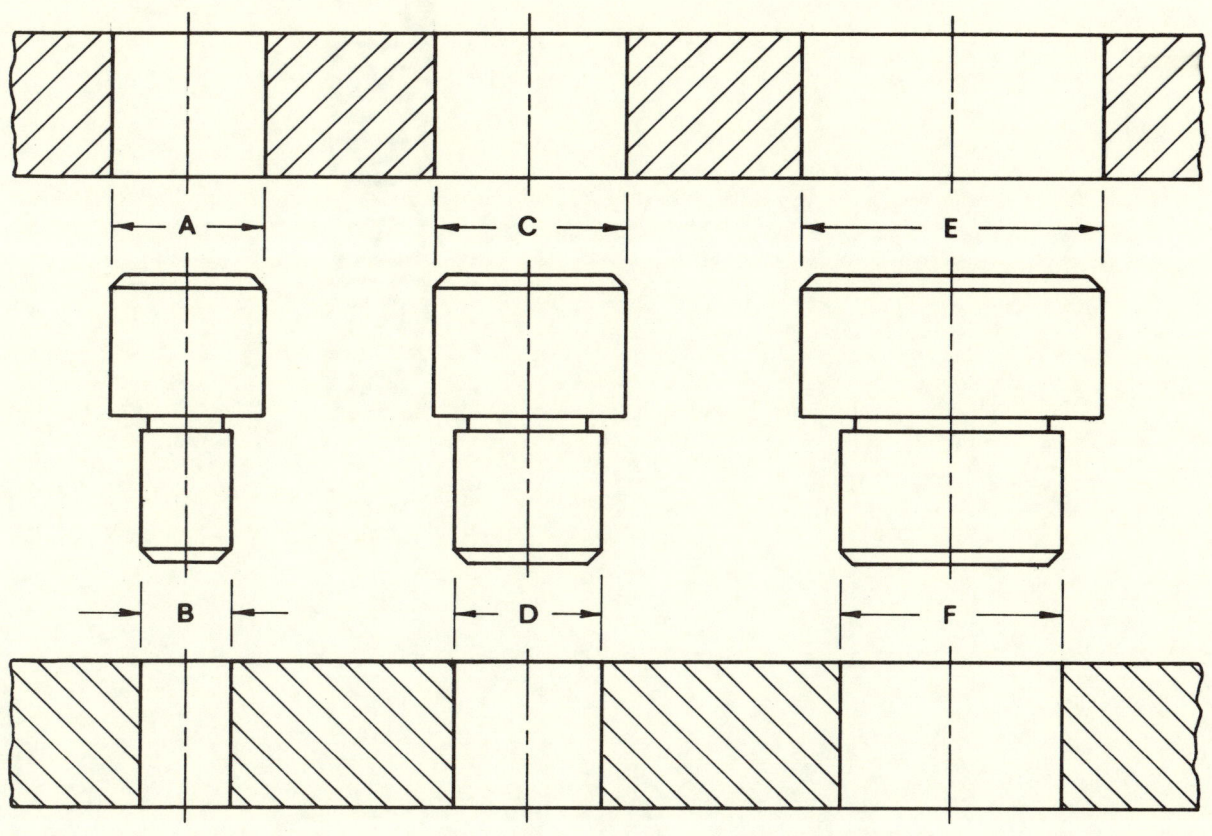

MATING UNIT	BASIC SIZE	ISO SYMBOL	BASIS	FEATURE	LIMITS OF SIZE		CLEARANCE OR INTERFERENCE	
					MAX	MIN	MAX	MIN
A	⌀20	H8/f7	HOLE	HOLE				
				SHAFT				
B	⌀12	H7/g6	HOLE	HOLE				
				SHAFT				
C	⌀25	K7/h6	SHAFT	HOLE				
				SHAFT				
D	⌀20	H7/p6	HOLE	HOLE				
				SHAFT				
E	⌀40	S7/h6	SHAFT	HOLE				
				SHAFT				
F	⌀30	H7/u6	HOLE	HOLE				
				SHAFT				

METRIC TOLERANCES

NAME
SECTION   DATE

GRADE 70

DIMENSION THE DRAWINGS BELOW INCORPORATING FEATURE CONTROL FRAMES

1
1. THE AXIS OF EACH HOLE MUST LIE WITHIN A TOLERANCE ZONE OF 0.02mm DIAMETER.
2. THE DATUM SURFACE A MUST BE FLAT WITHIN 0.02mm.
3. SURFACE B MUST BE PARALLEL TO DATUM SURFACE A WITHIN 0.08mm.

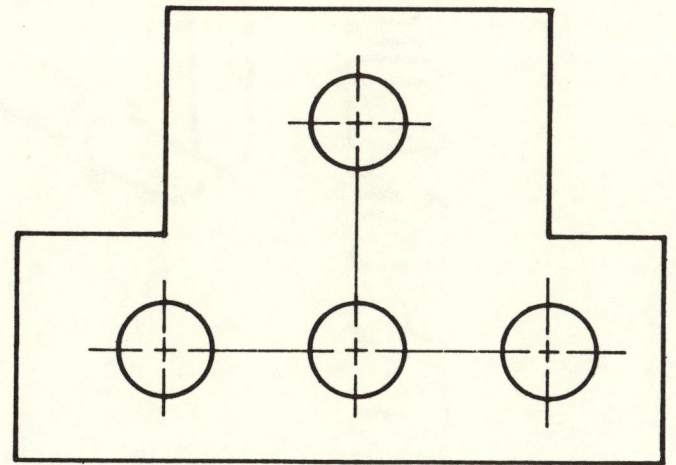 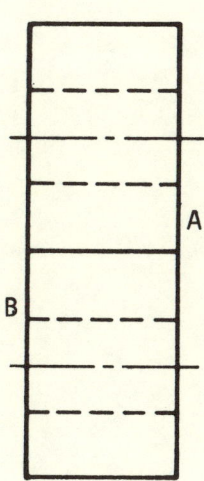

2
1. CYLINDERS A AND B ARE TO HAVE A CYLINDRICITY TOLERANCE OF 0.05mm. CYLINDER C IS THE DATUM.
2. THE SMALL BORE MUST BE CONCENTRIC WITH CYLINDER C WITHIN 0.015mm.

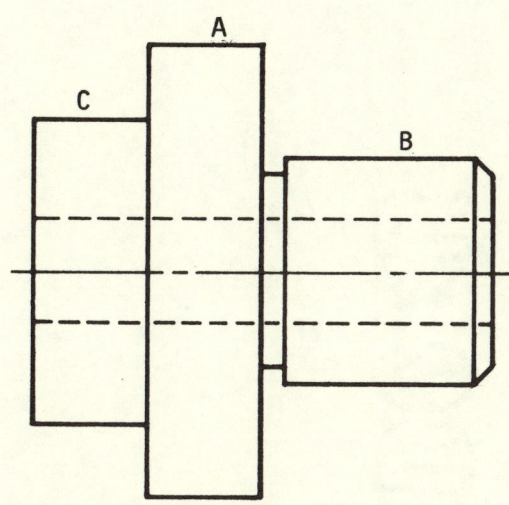 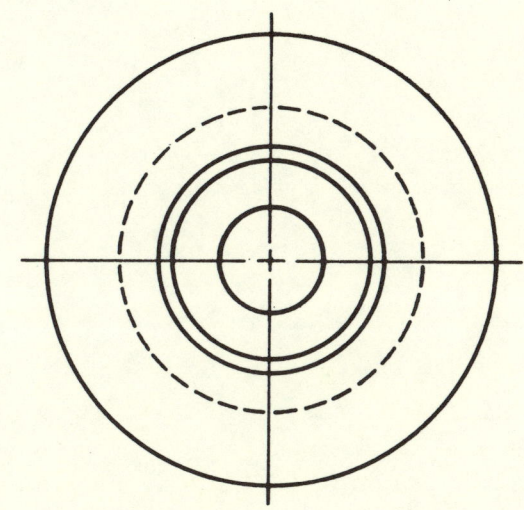

GEOMETRIC TOLERANCING

DIMENSION THE DRAWINGS BELOW INCORPORATING FEATURE CONTROL FRAMES

1
1. THE VERTICAL BORE B MUST BE PERPENDICULAR TO DATUM BORE A WITHIN A TOLERANCE OF 0.03mm.
2. SURFACES C AND D MUST BE PERPENDICULAR TO DATUM SURFACE E WITHIN 0.02mm.

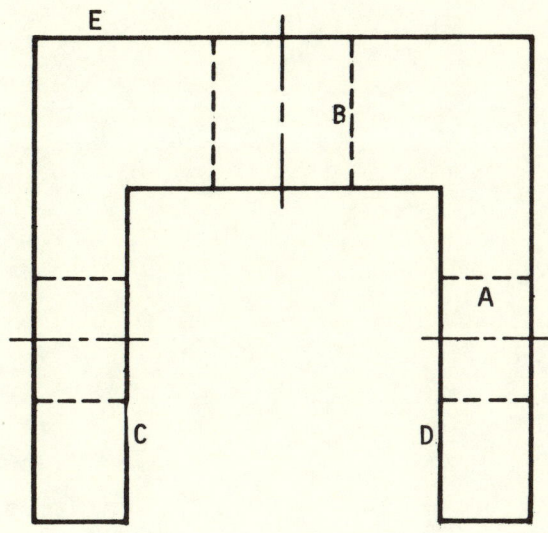
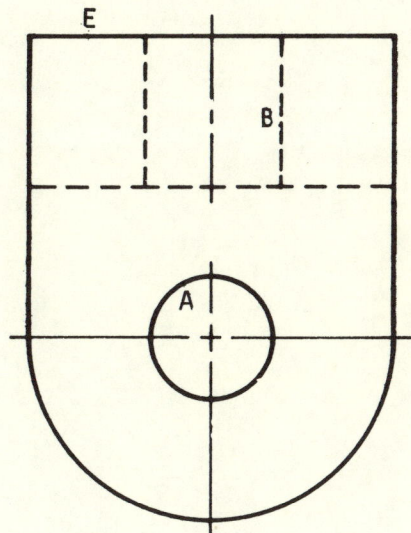

2
1. THE AXIS OF THE BORE MUST BE PARALLEL TO DATUM SURFACE A WITHIN 0.07mm.
2. SURFACES B AND C MUST BE PERPENDICULAR TO DATUM SURFACE A WITHIN 0.02mm.

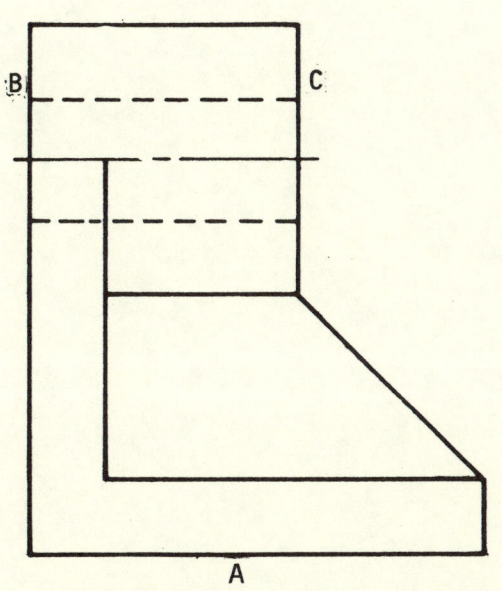
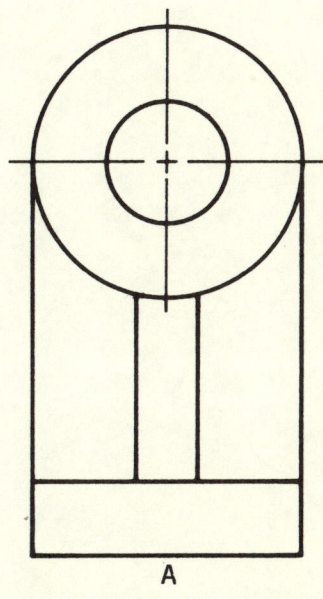

GEOMETRIC TOLERANCING | NAME | SECTION | DATE | GRADE | 72

DIMENSION THE DRAWING BELOW INCORPORATING FEATURE CONTROL FRAMES.

1. THE DATUM SURFACES A AND B MUST BE FLAT WITHIN A TOLERANCE ZONE OF 0.02mm.
2. THE SURFACE C MUST BE PARALLEL TO DATUM SURFACES A AND B WITHIN 0.08mm.
3. THE LARGE BORE MUST BE PERPENDICULAR TO DATUM SURFACES A AND B WITHIN 0.08mm.
4. THE LARGE BORE MUST HAVE A CYLINDRICITY TOLERANCE OF 0.05mm.
5. THE AXES OF BOTH SMALL BORES MUST BE PARALLEL TO THE LARGE BORE WITHIN 0.04mm.
6. BOTH SMALL BORES MUST BE STRAIGHT WITHIN 0.03mm.
7. THE LARGE HUB MUST BE CIRCULAR WITHIN A TOLERANCE ZONE OF 0.02mm AT MMC.
8. THE SIDES OF THE 4X8 KEYWAY ARE TO BE GIVEN A SURFACE-TEXTURE SYMBOL, WITH ROUGHNESS APPROPRIATE TO A MILLING OPERATION.

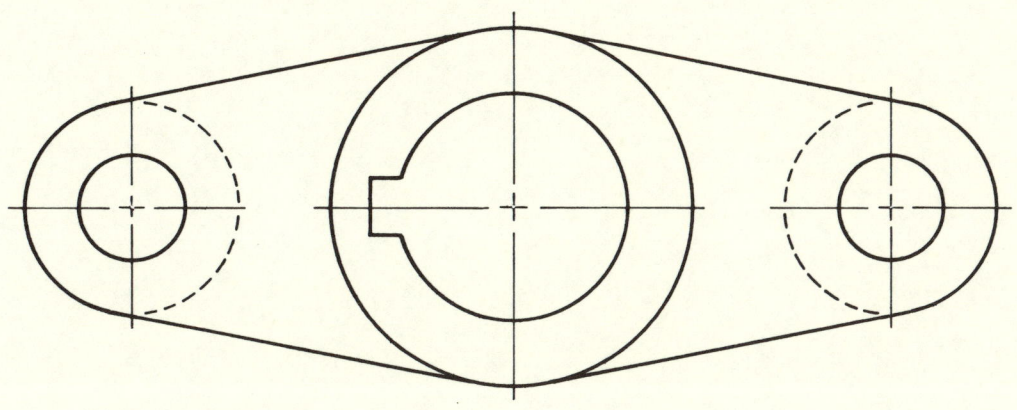

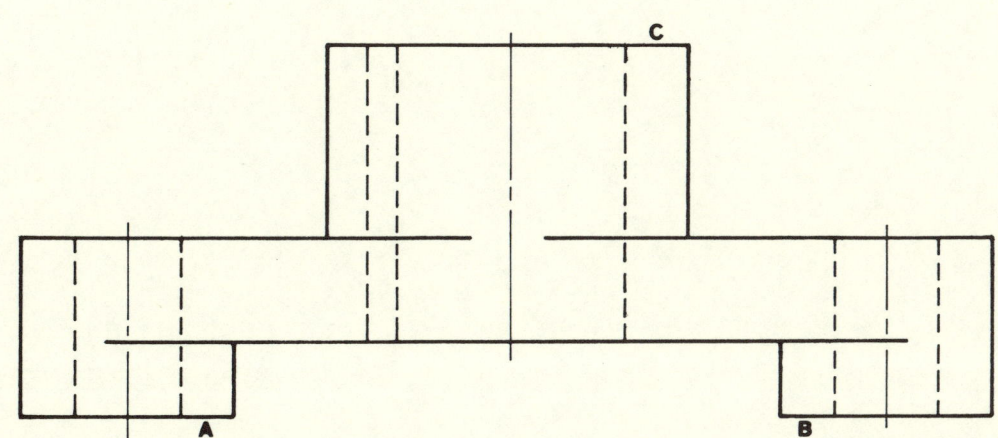

GEOMETRIC TOLERANCING

THE PLANES OF PROJECTION AND MEASUREMENTS

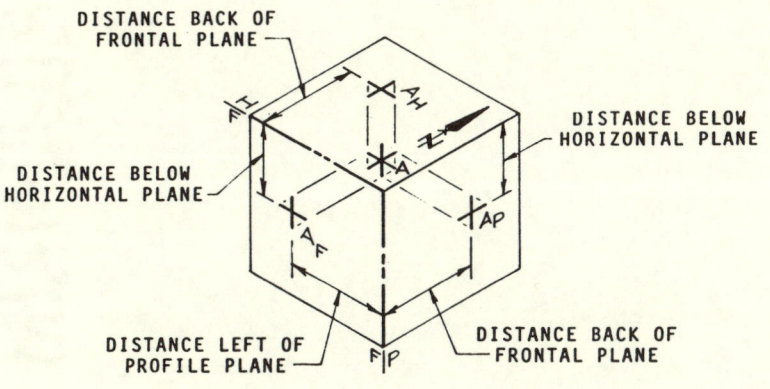

1 CONSIDER THE PLANES OF PROJECTION AS THE CEILING, FRONT AND SIDE WALLS OF YOUR ROOM.

LOCATE THE FOLLOWING:

1. A TELEPHONE OUTLET ON THE SIDE WALL, 2'-6" BEHIND THE FRONT WALL AND 5'-3" BELOW THE CEILING.
2. A LIGHT FIXTURE ON THE CEILING, 4'-0" LEFT OF THE SIDE WALL AND 3'-0" BEHIND THE FRONT WALL.
3. AN ELECTRICAL OUTLET ON THE FRONT WALL, 4'-6" TO THE LEFT OF THE SIDE WALL AND 6'-0" BELOW THE CEILING.

SCALE: 3/8" = 1'-0"

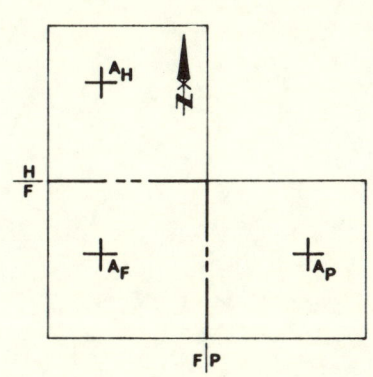

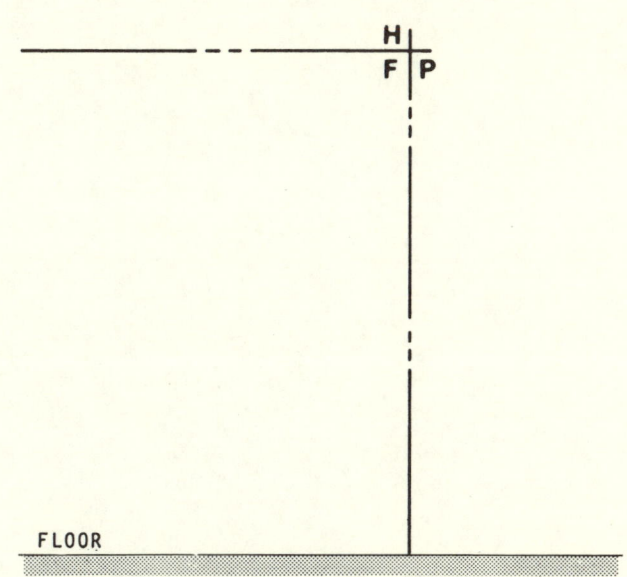

2 DRAW THE FOLLOWING:

"B" - 15mm LOWER THAN, 15mm EAST OF AND 25mm NORTH OF POINT A.

"C" - 30mm DUE WEST OF AND 25mm BELOW POINT A.

WITH A-B AND B-C AS TWO ADJACENT EDGES OF A PARALLELOGRAM DRAW THE PARALLELOGRAM IN ALL THREE VIEWS.

| SPACE GEOMETRY POINTS & LINES | NAME SECTION DATE | GRADE 74 |

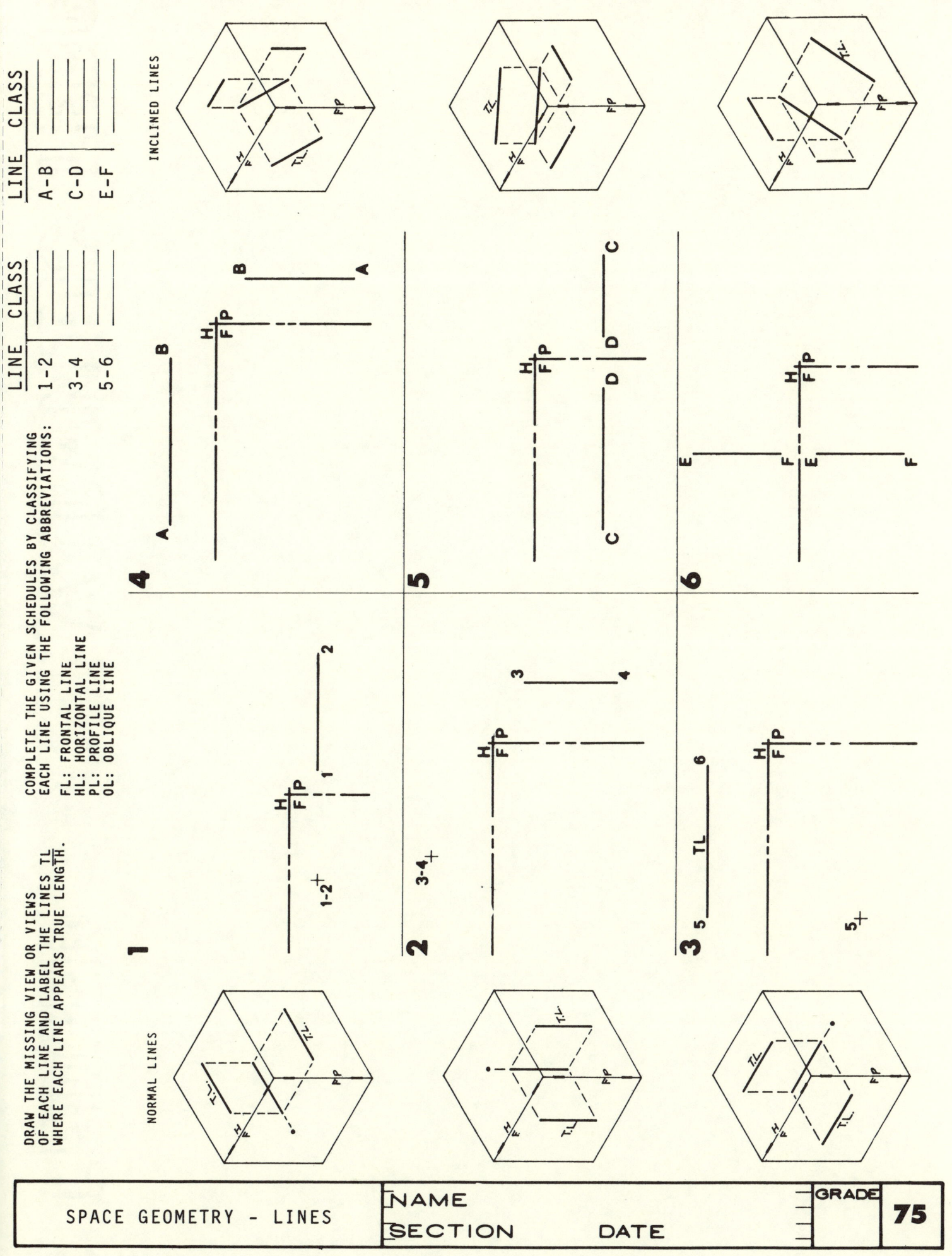

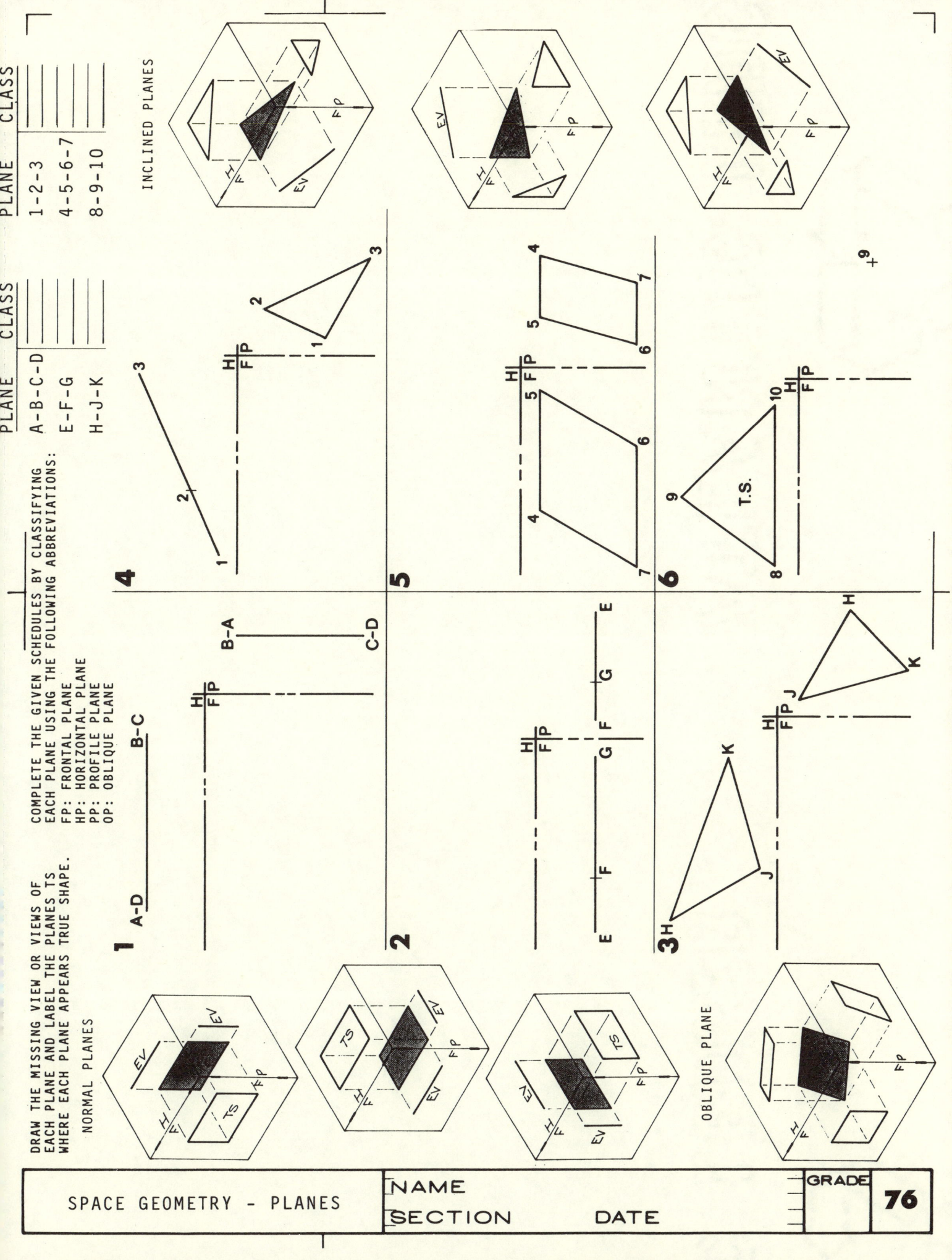

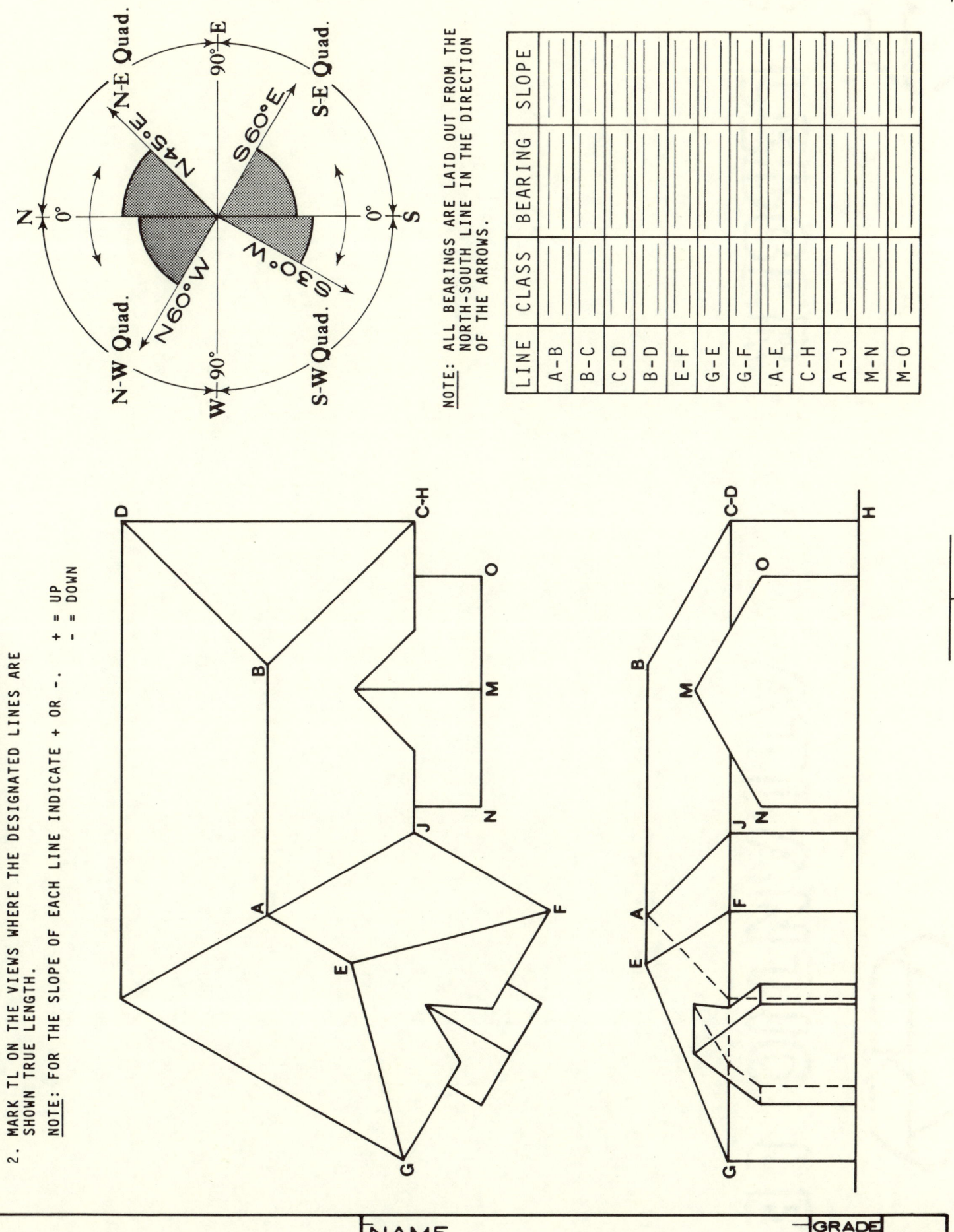

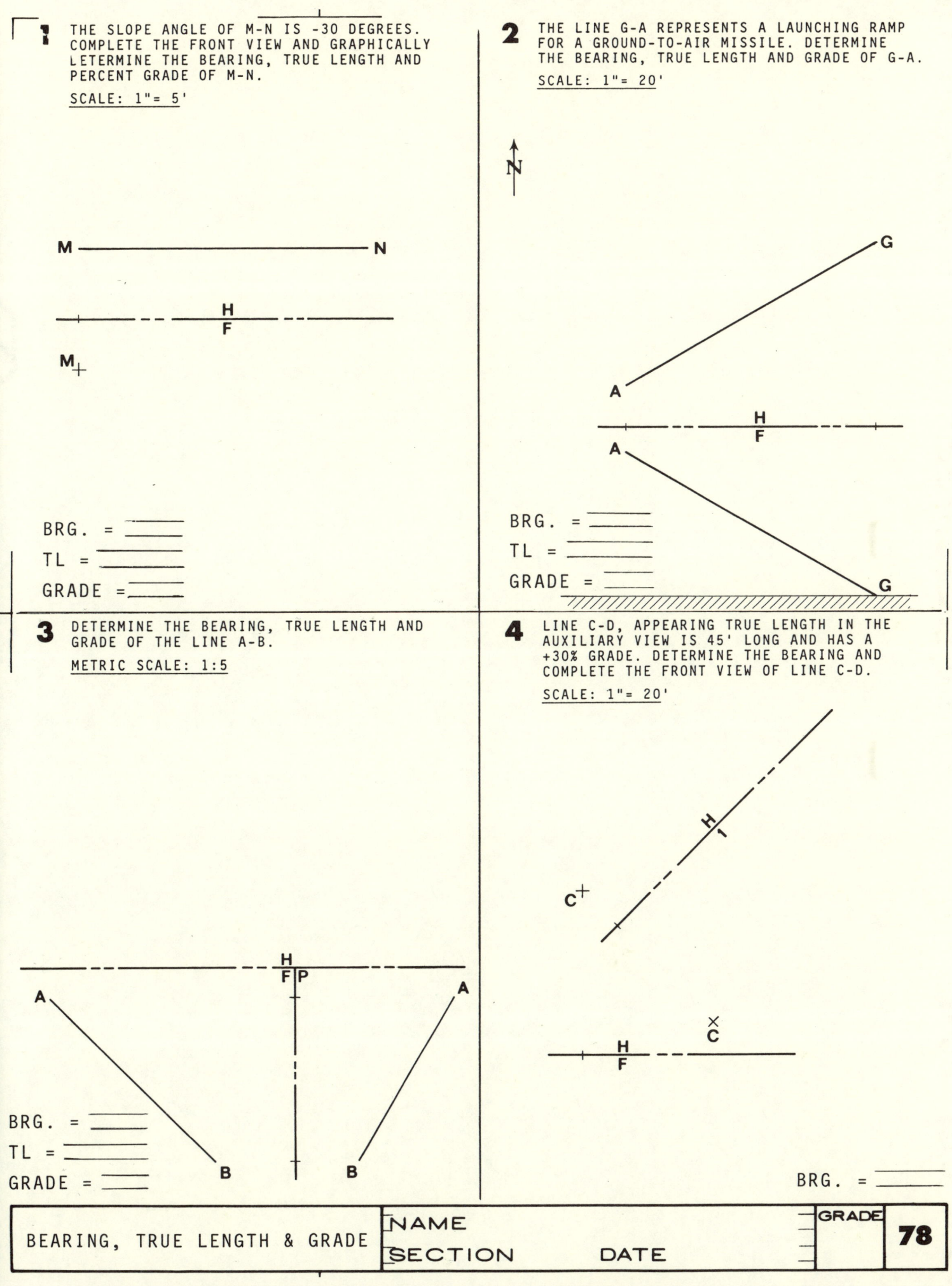

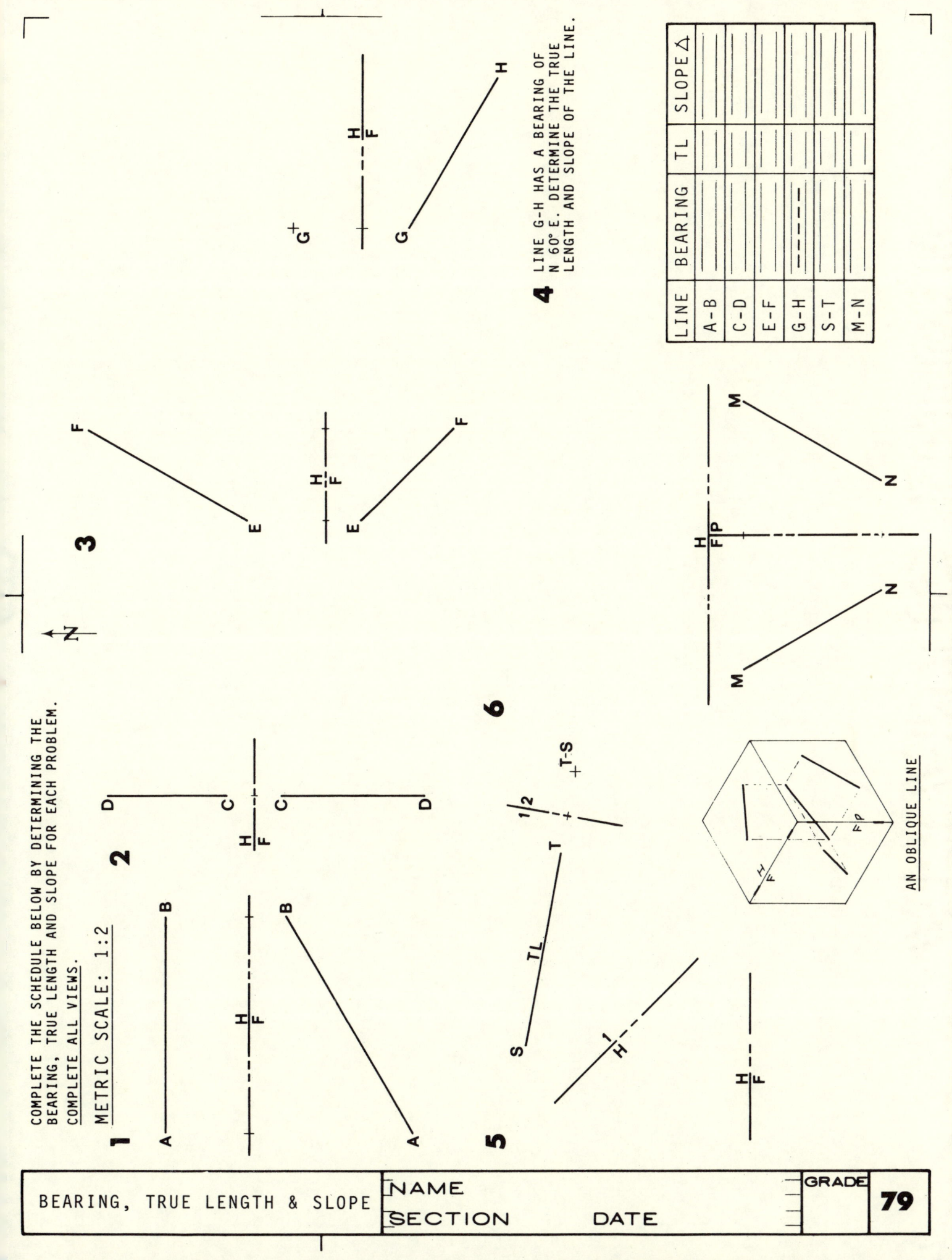

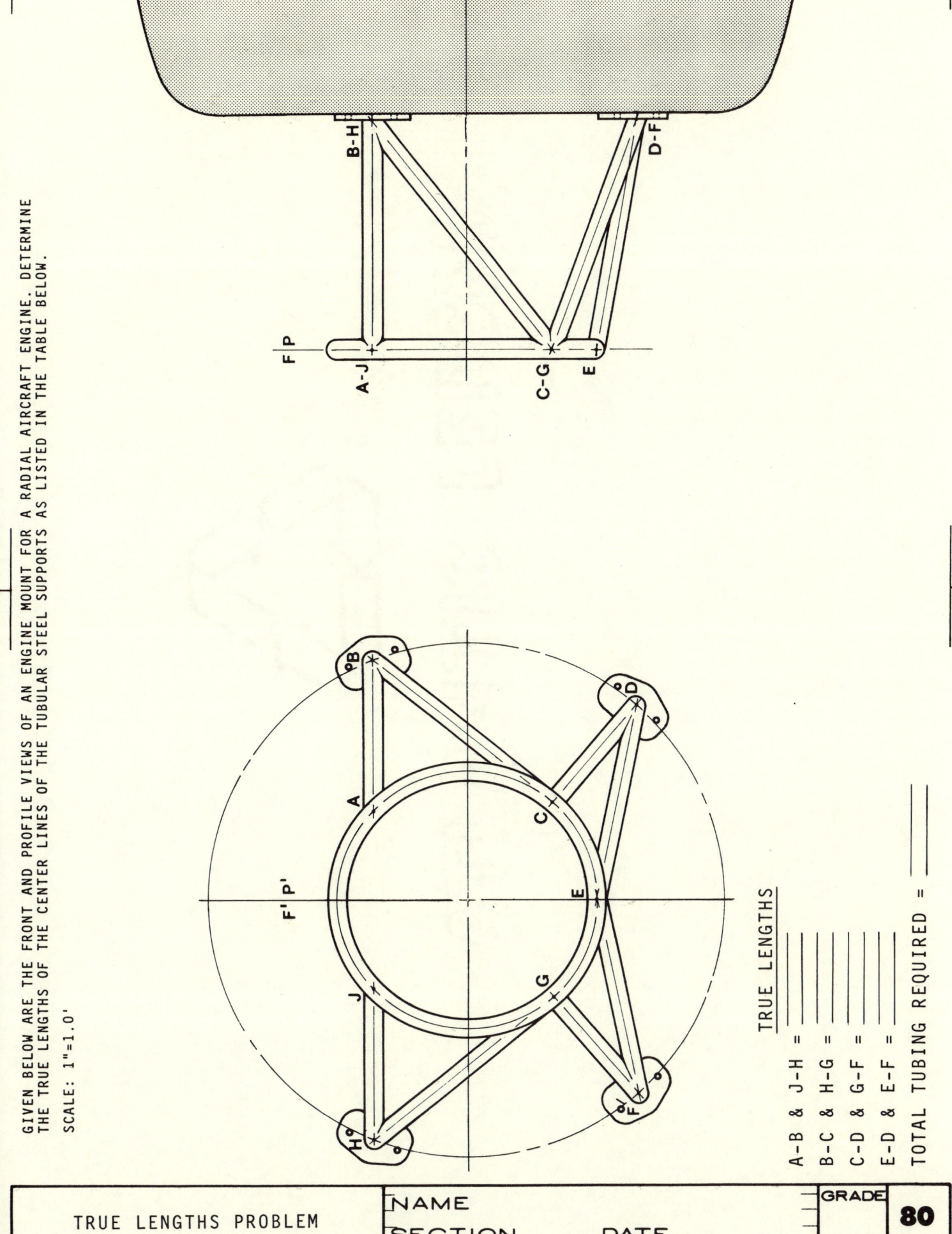

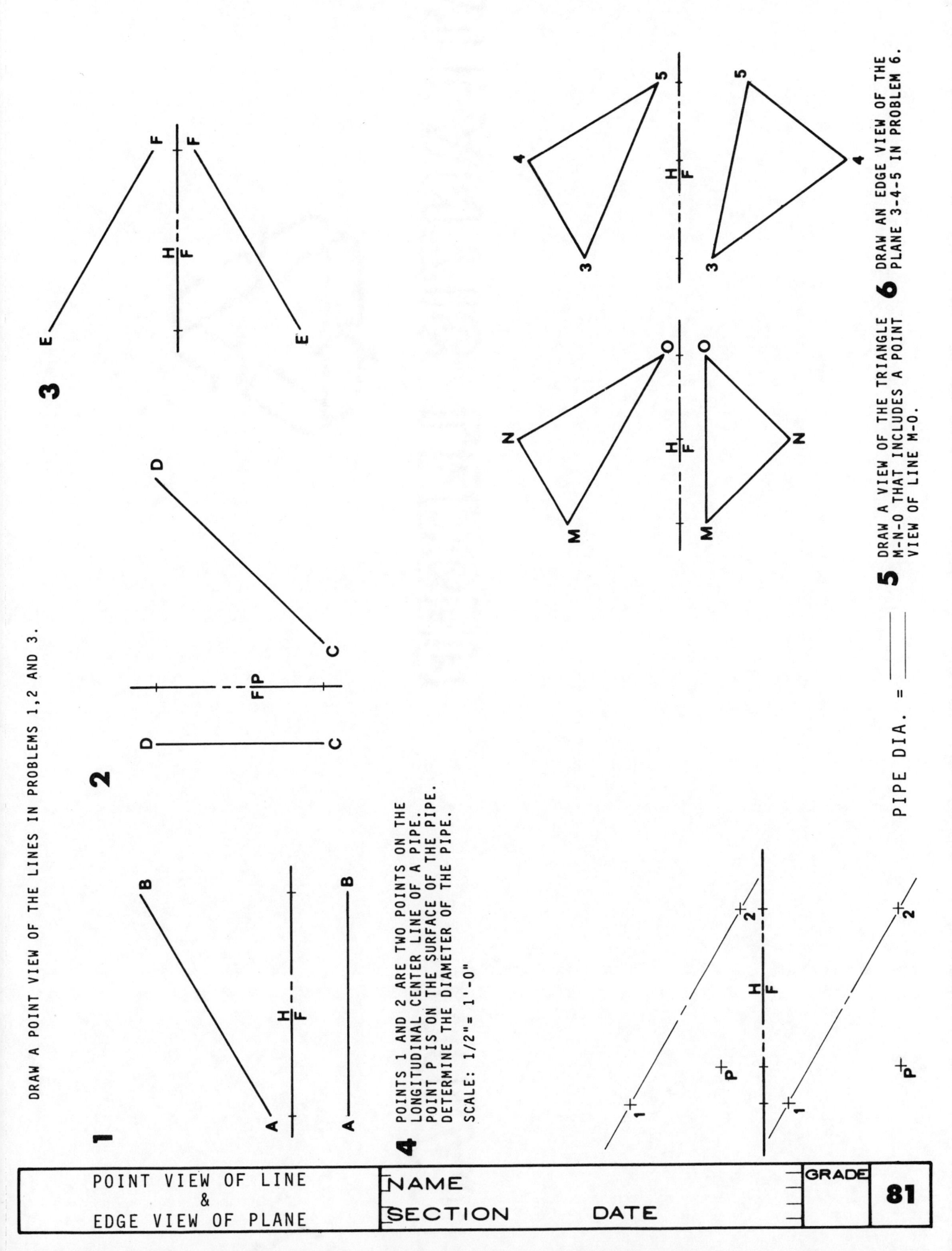

DRAW THE TRUE-SIZE VIEWS OF THE PLANES
FOR ALL OF THE PROBLEMS SHOWN BELOW.

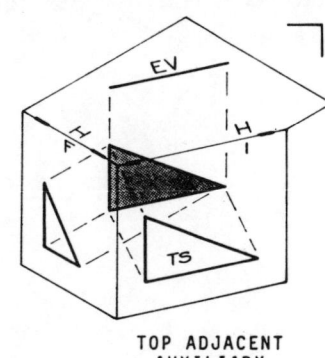

TOP ADJACENT
AUXILIARY

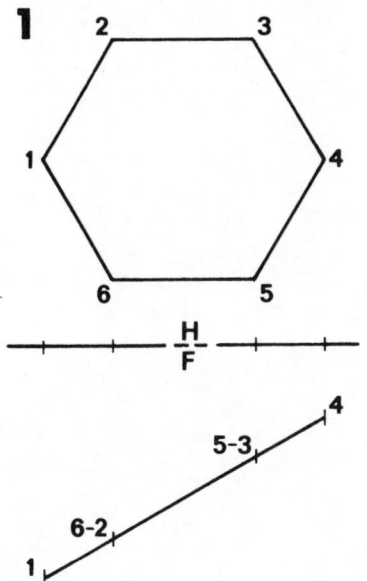

1

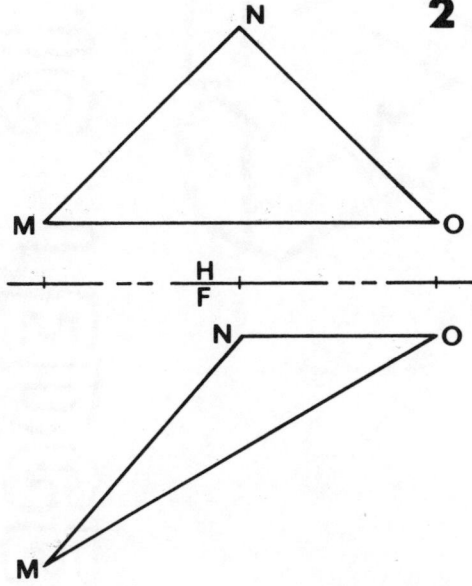

2

OPTIONAL: COMPUTE THE AREAS OF
THE TRUE-SIZE PLANES FOR
PROBLEMS 2 AND 3.
METRIC SCALE: 1:1000

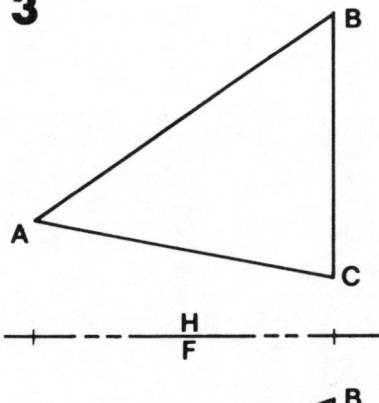

3

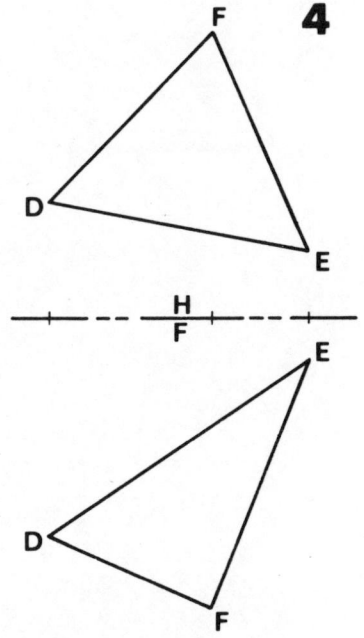

4

AREAS

NO. 2 = _____ M^2
NO. 3 = _____ M^2

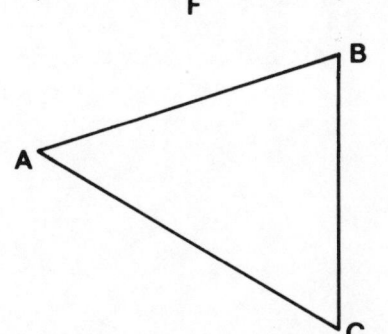

TRUE SIZE OF PLANES

NAME
SECTION   DATE

GRADE
82

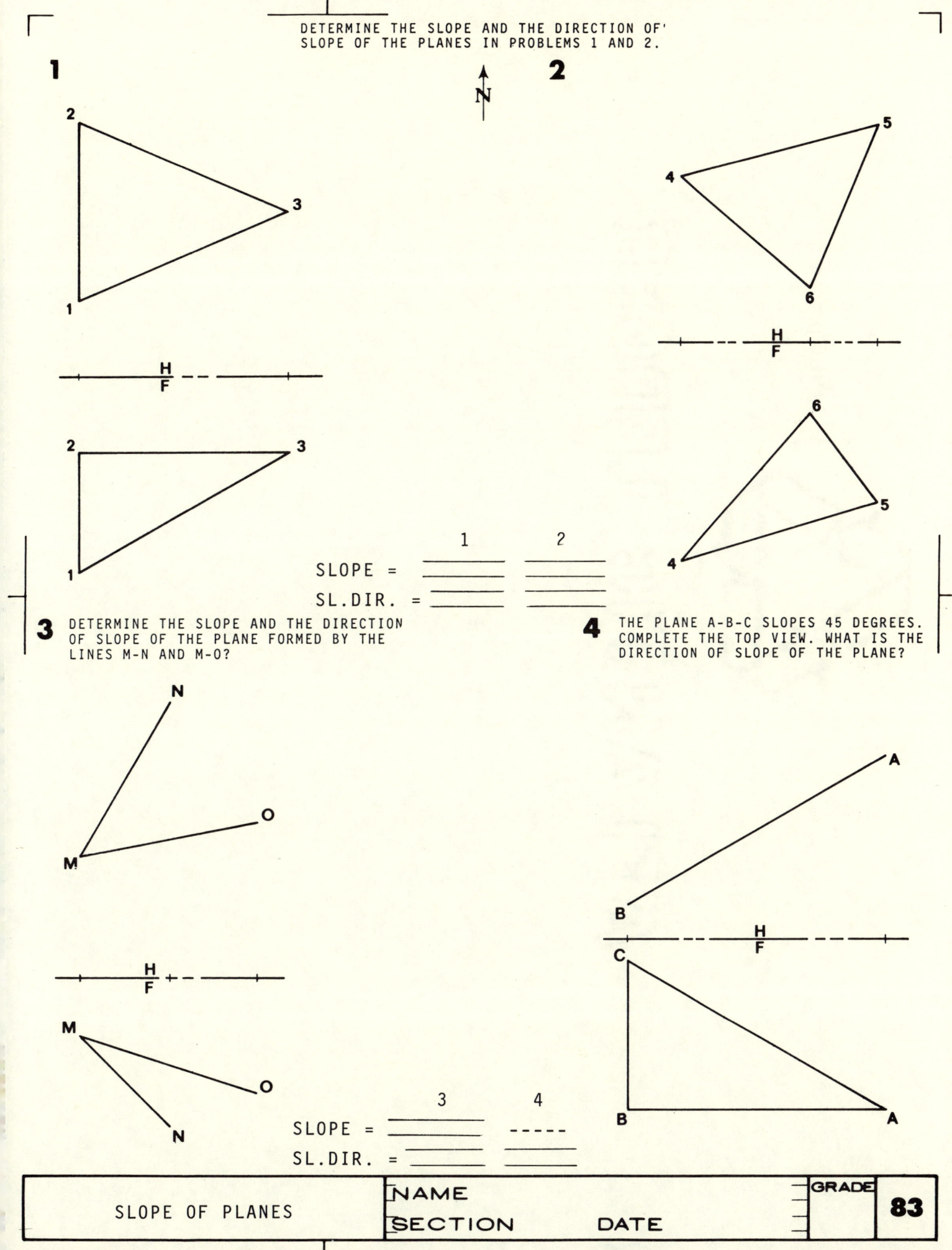

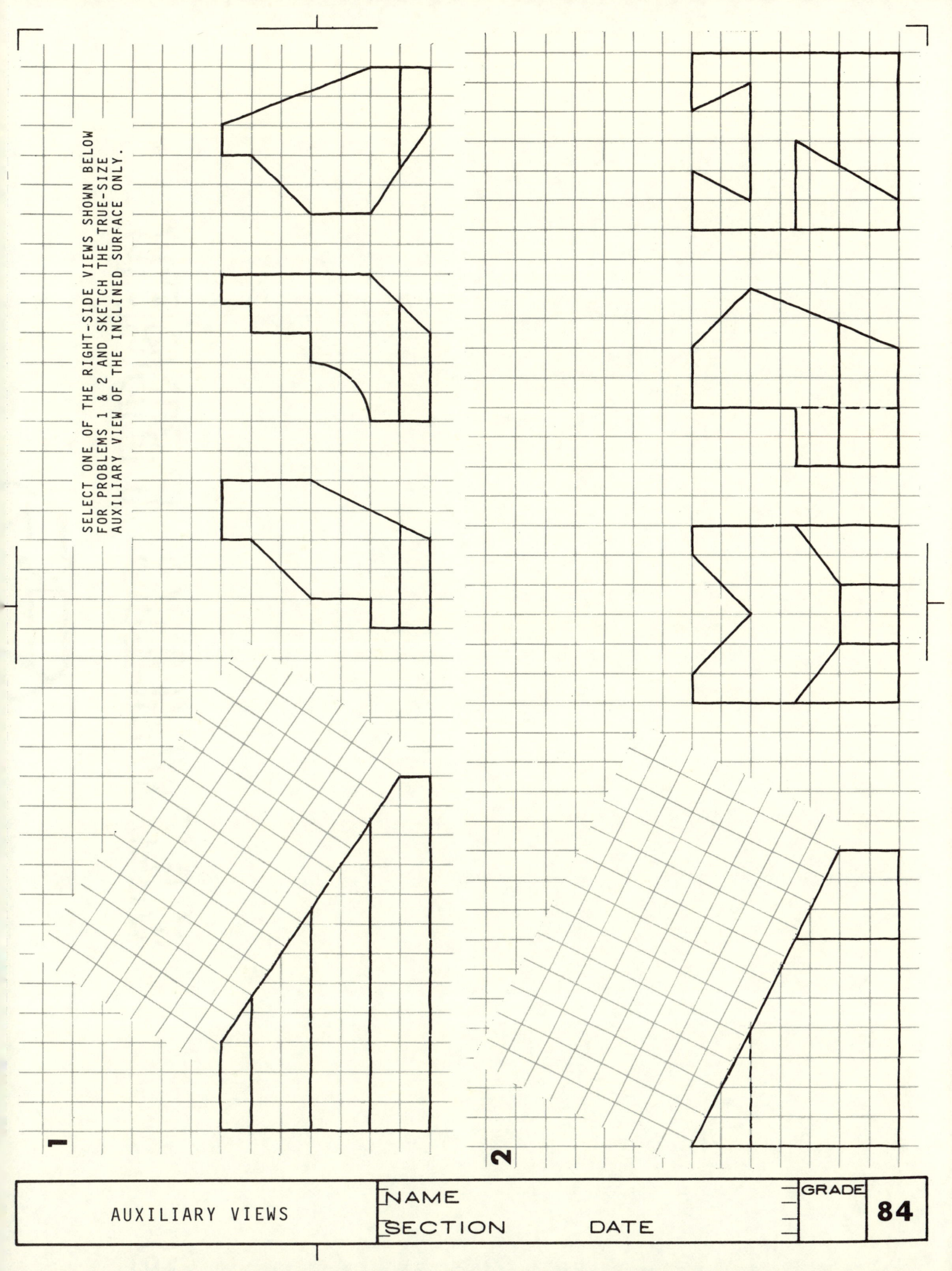

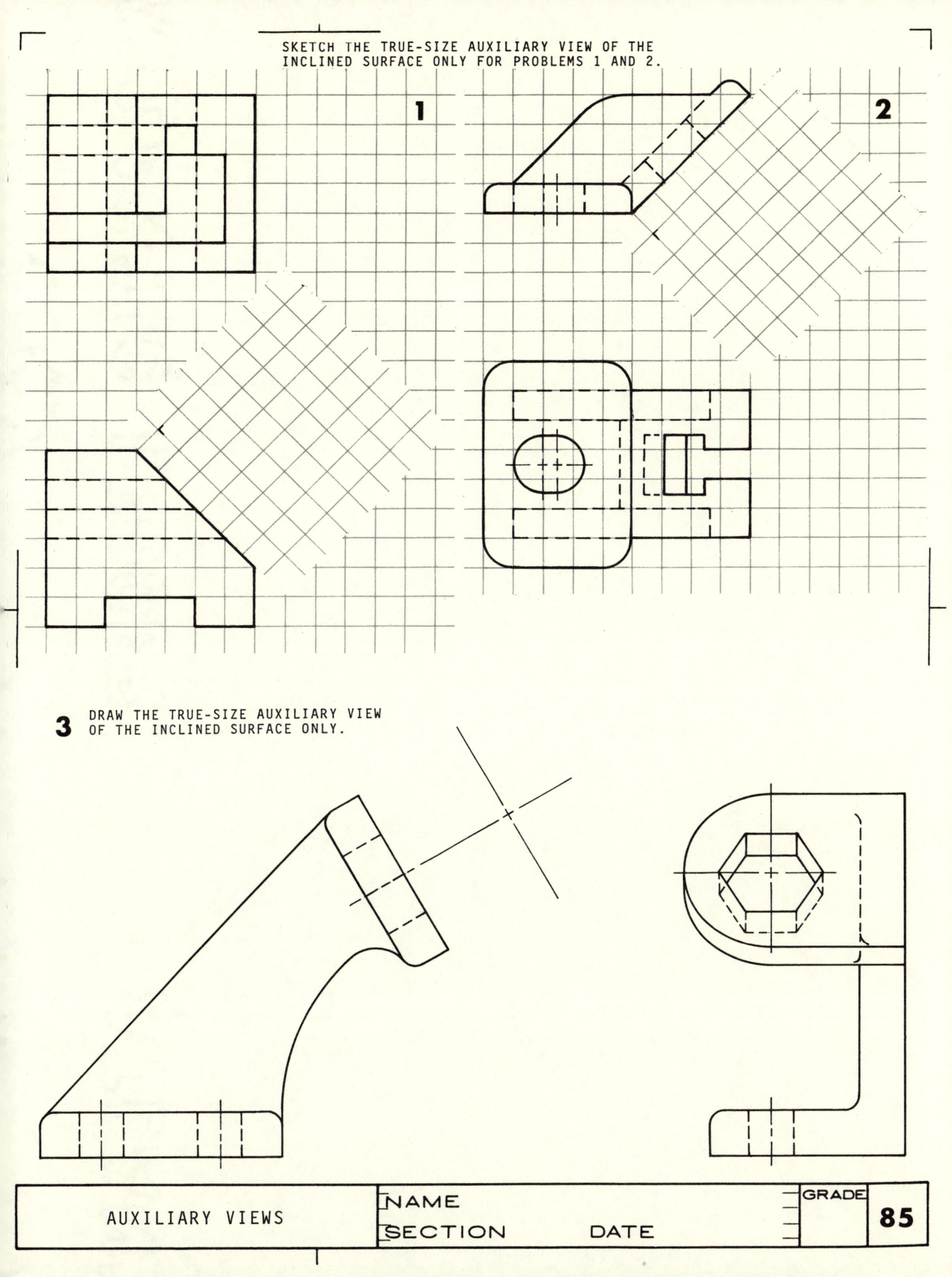

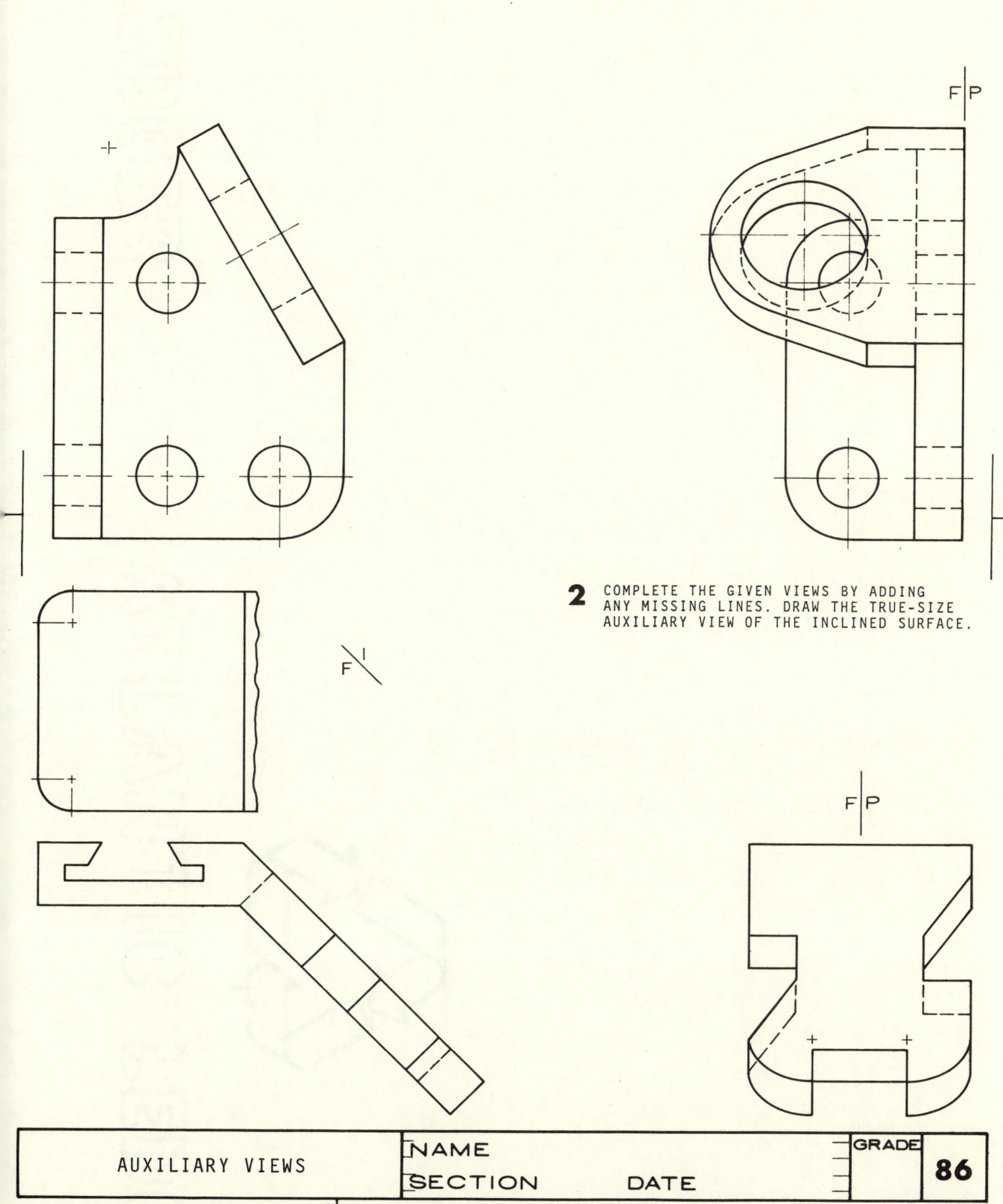

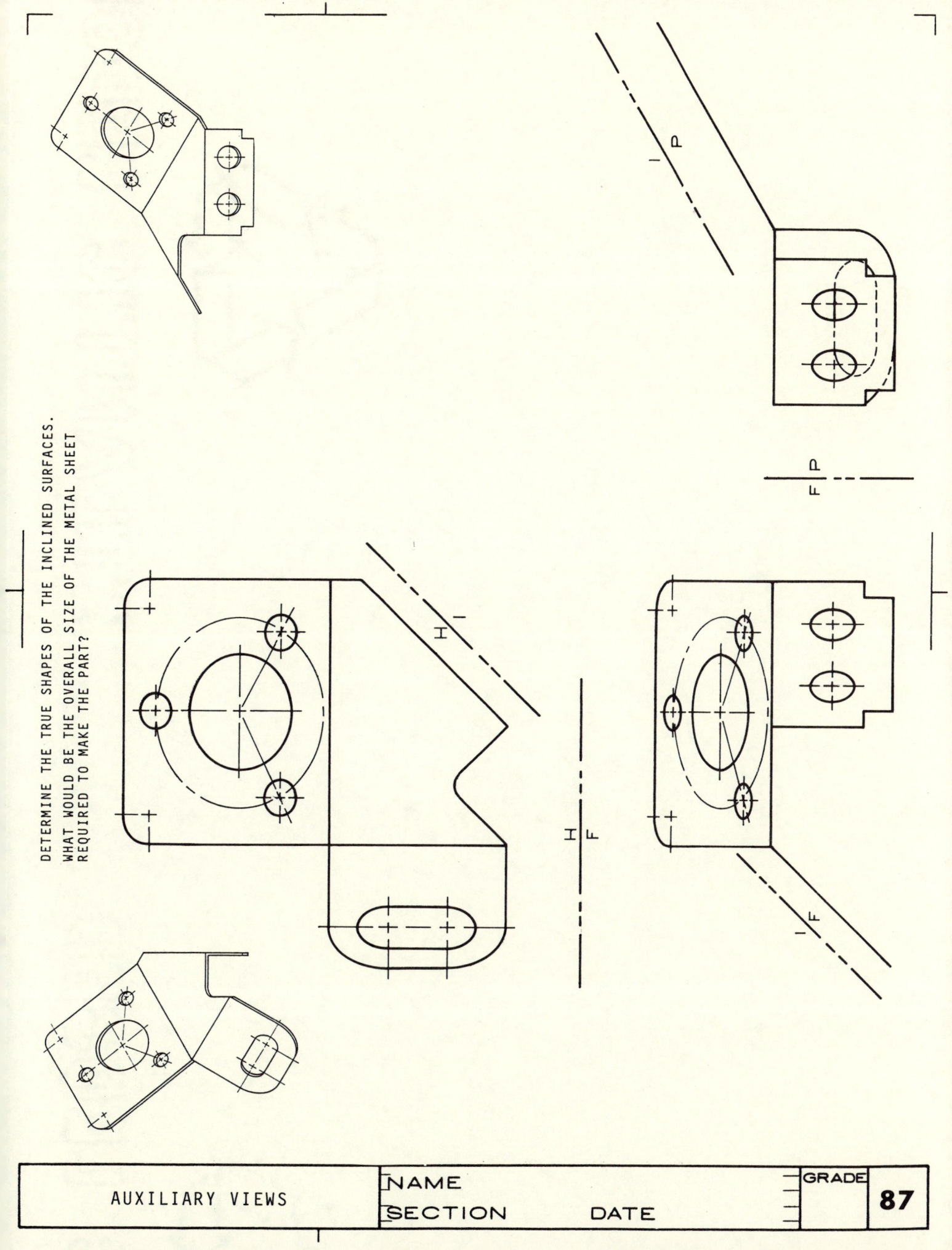

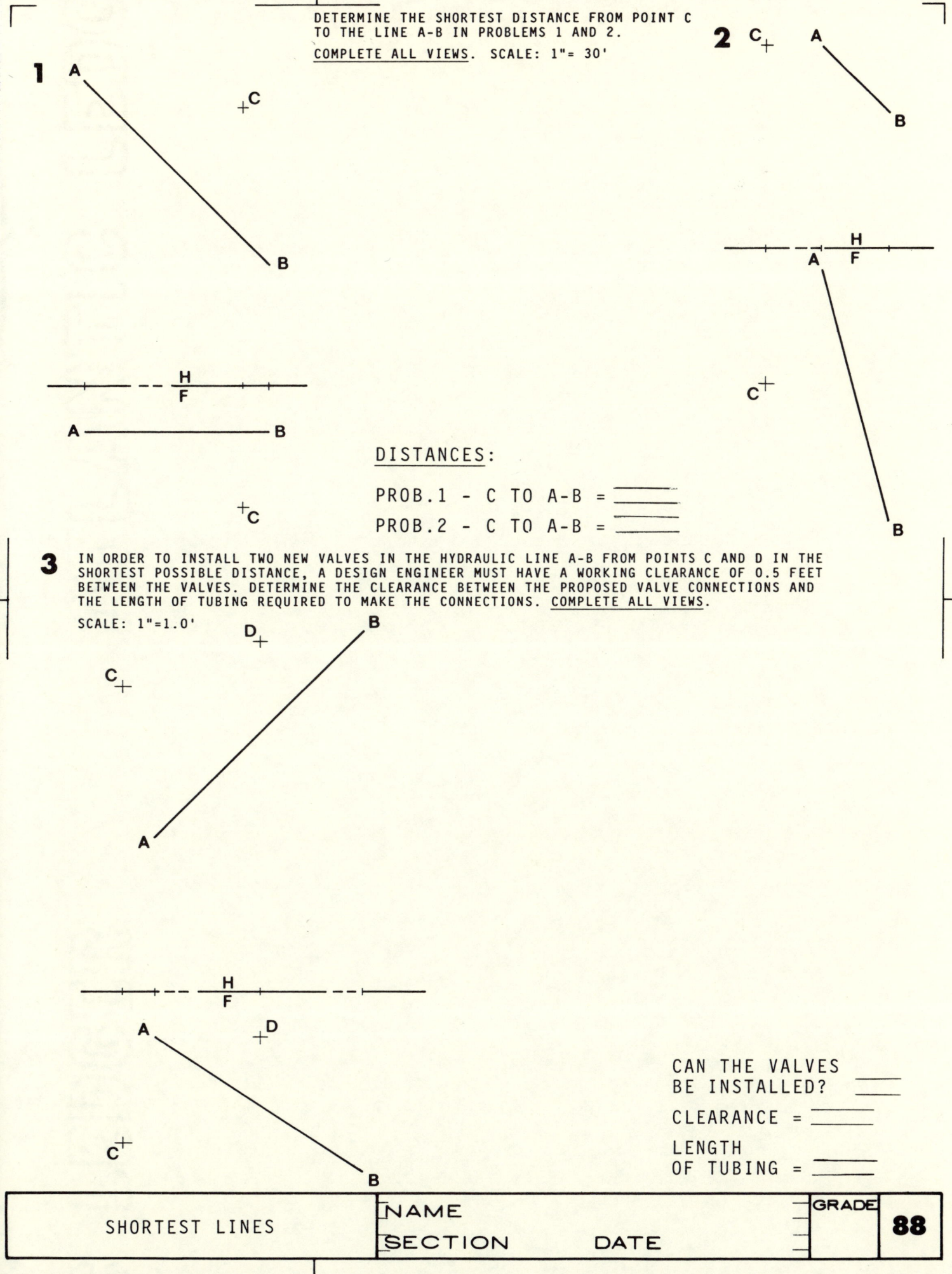

1 DRAW AND MEASURE THE SHORTEST DISTANCE (CLEARANCE) AND THE SHORTEST VERTICAL DISTANCE BETWEEN THE RIGID POWER LINE E-L AND THE GUY WIRE G-W. SHOW THE DISTANCES IN ALL VIEWS.

SCALE: 1/16"= 1'-0"

2 DRAW AND MEASURE THE SHORTEST DISTANCE AND THE SHORTEST HORIZONTAL DISTANCE BETWEEN THE LINES A-B AND C-D. SHOW THE DISTANCES IN ALL VIEWS.

METRIC SCALE: 1:5

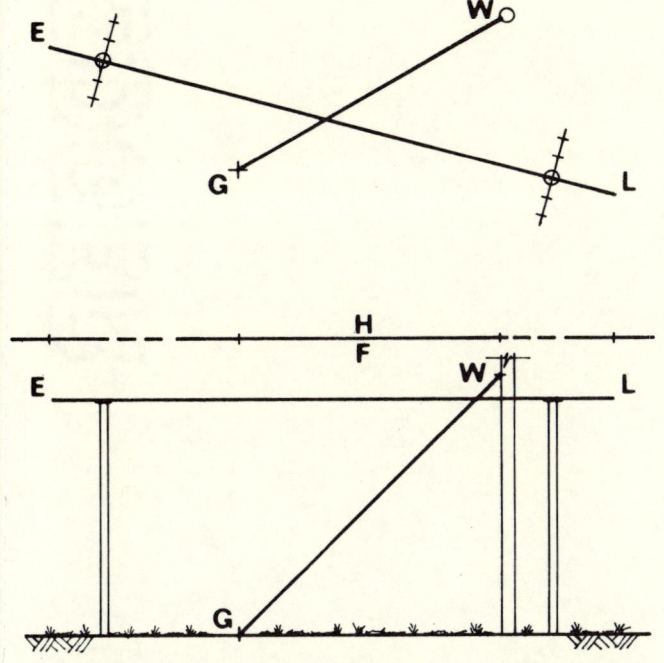
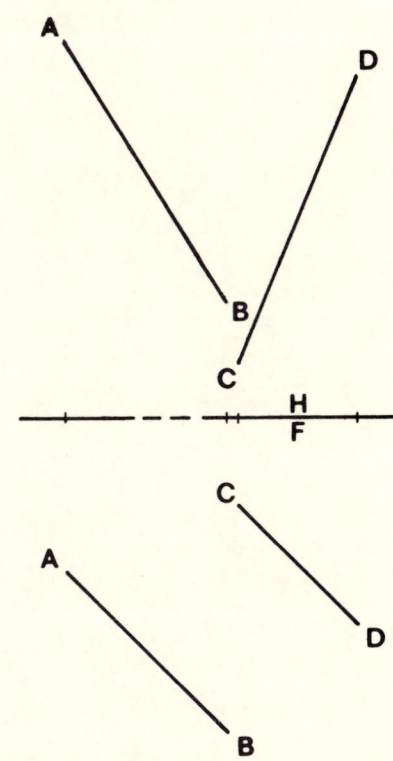

3 DRAW AND MEASURE THE SHORTEST HORIZONTAL DISTANCE AND THE SHORTEST VERTICAL DISTANCE BETWEEN THE LINES M-N AND O-P. SHOW THE DISTANCES IN THE APPROPRIATE VIEWS.

METRIC SCALE: 1:1

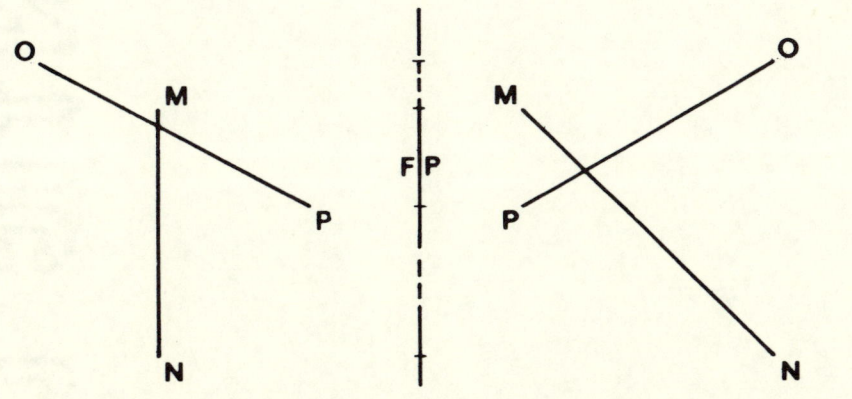

ANSWERS

	S.DIST.	S.V.DIST.
PROB.1 -		

	S.DIST.	S.H.DIST.
PROB.2 -		

	S.H.DIST.	S.V.DIST.
PROB.3 -		

SKEWED LINES | NAME SECTION DATE | GRADE **89**

1 DRAW AND MEASURE THE TRUE LENGTH OF THE SHORTEST LINE FROM POINT D TO THE PLANE A-B-C FOR PROBLEMS 1 AND 2. DRAW THE LINE IN ALL OF THE VIEWS. METRIC SCALE: 1:2 **2**

PROB.1 - TL= _____
PROB.2 - TL= _____

3 THE PLANE FORMED BY THE POINTS 1-2-3-4 IS THE BASE OF A RIGHT PYRAMID WITH VERTEX V AT AN ALTITUDE OF 1.5" ABOVE THE BASE AT ITS MIDPOINT. DRAW THE PYRAMID IN ALL VIEWS SHOWING THE CORRECT VISIBILITY. ANSWER THE QUESTIONS LISTED BELOW. SCALE: 1"= 1"

$V = \frac{1}{3} A_B H$ A_B = AREA OF THE BASE

1. IS THE BASE RECTANGULAR? _____
2. WHAT IS ITS AREA? _____
3. WHAT IS THE VOLUME OF THE PYRAMID? _____

PERPENDICULARS

NAME
SECTION   DATE
GRADE
90

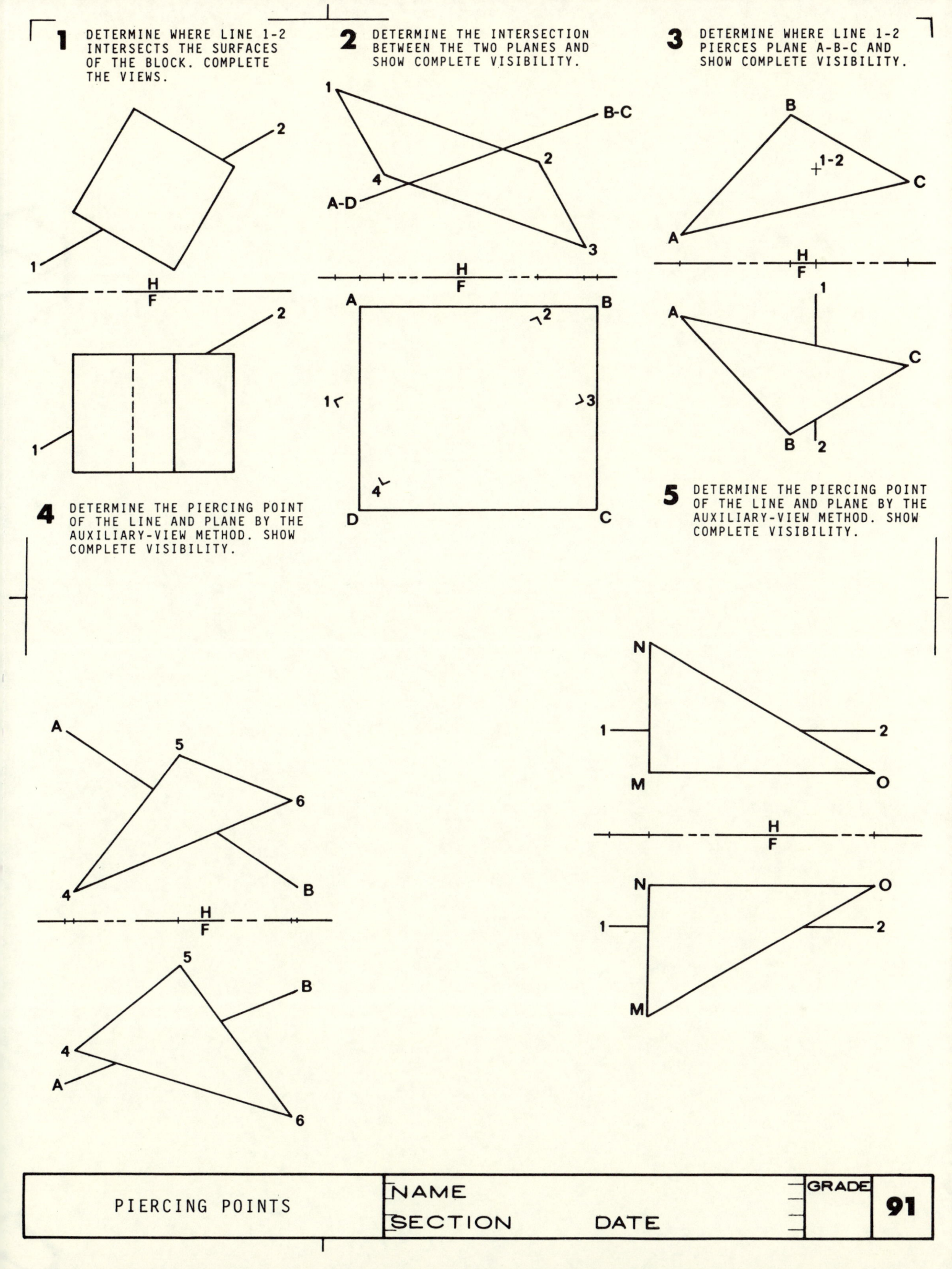

1 FLAGPOLE A-B IS SECURED BY TWO BRACES FROM POINTS C AND D ON THE FACE OF THE WALL. DETERMINE THEIR TRUE LENGTHS AND THE ANGLE FORMED BETWEEN THE BRACES. WHAT IS THE TRUE LENGTH OF THE FLAGPOLE? WHAT IS THE ANGLE FORMED BETWEEN THE BRACES AND THE FLAGPOLE?
SCALE: 1/4"=1'-0"

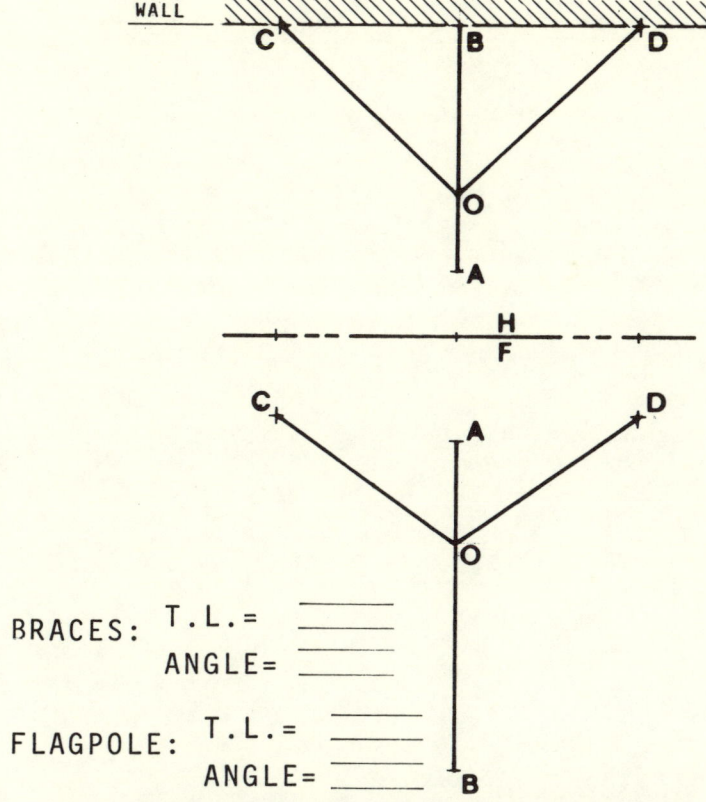

BRACES: T.L. = _____
ANGLE = _____

FLAGPOLE: T.L. = _____
ANGLE = _____

2 A GENERAL ELECTRIC SPACE STATION PLATFORM IS TO BE CONSTRUCTED OF ROUND TUBULAR ALUMINUM COMPONENTS WITH A 4" OUTSIDE DIAMETER. TO SIMPLIFY THE MILLING AND WELDING OF THE ENDS OF THE TUBES, SPHERICAL ALUMINUM CONNECTORS HAVE BEEN SPECIFIED. WHAT IS THE MINIMUM DIAMETER SPHERICAL CONNECTOR WHICH MAY BE SPECIFIED FOR THE JOINT INDICATED BELOW?

SUGGESTED SCALE: 3"= 1'-0" TO DETERMINE THE SPHERICAL DIAMETER

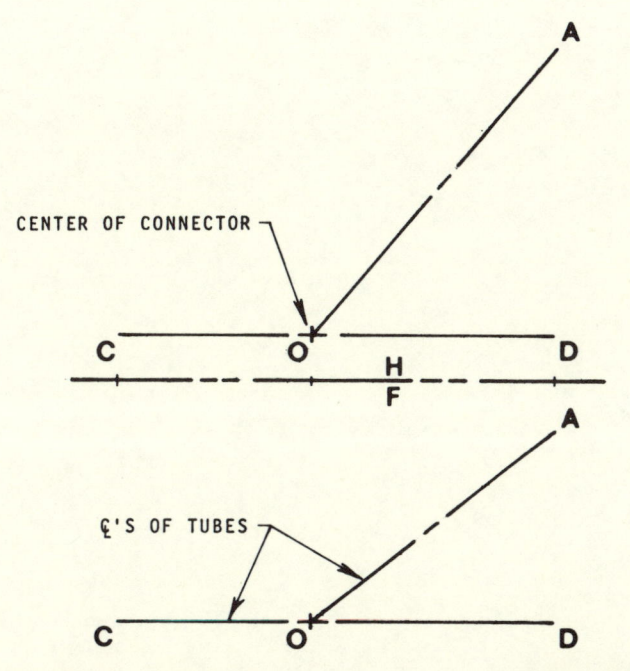

SPHERICAL DIAMETER = _____

ANGLES BETWEEN LINES | 92

DETERMINE THE ANGLES BETWEEN THE PLANES IN THE PROBLEMS BELOW BY THE AUXILIARY VIEW METHOD.

1

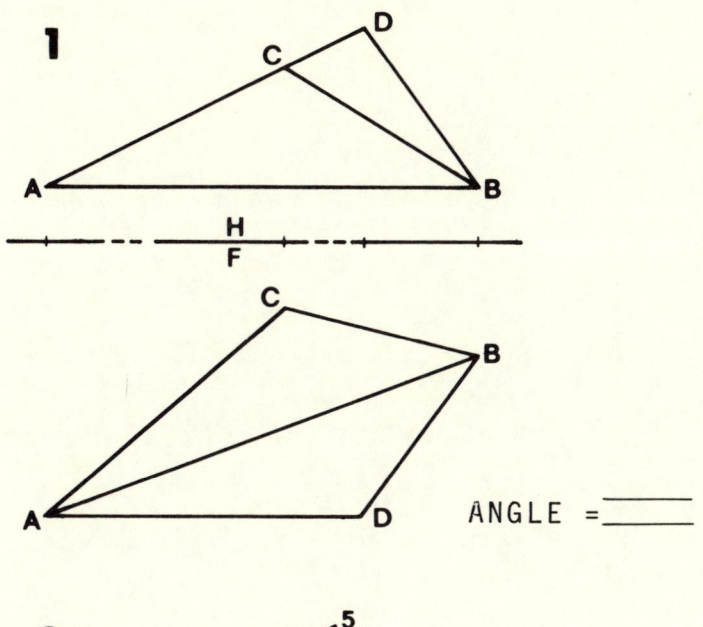

ANGLE = ____

2

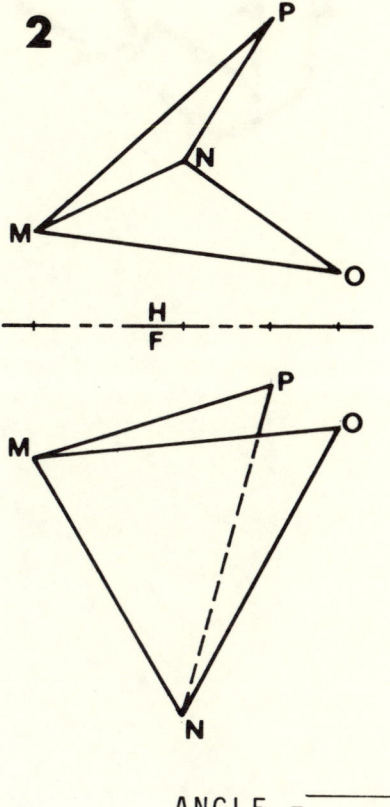

ANGLE = ____

3

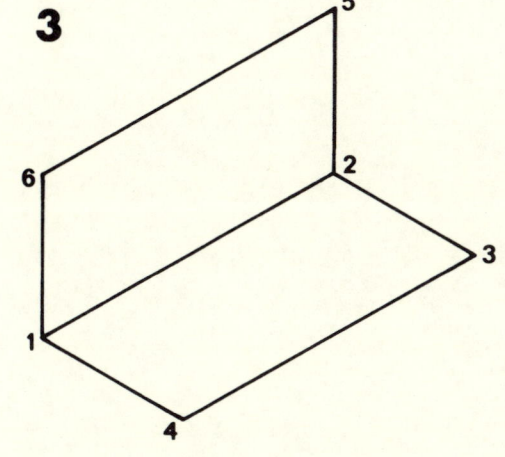

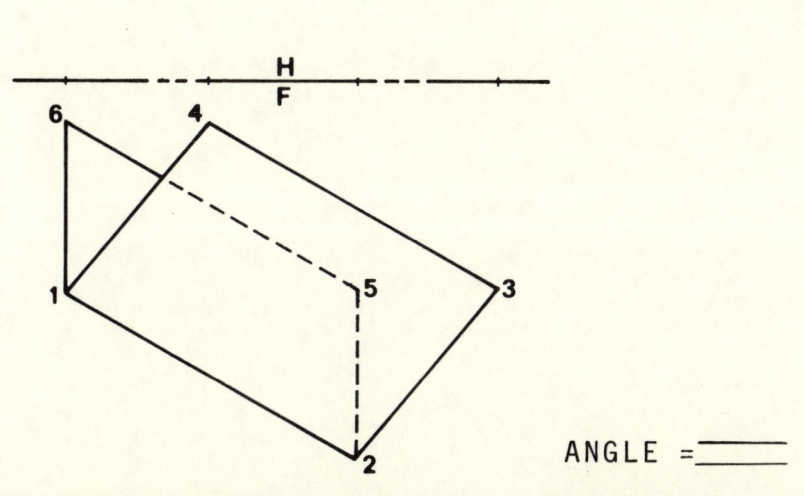

ANGLE = ____

ANGLES BETWEEN PLANES	NAME		GRADE	93
	SECTION	DATE		

1 DETERMINE THE LOCATION OF POINT 1 WHERE IT INTERSECTS THE PLANE A-B-C-D. WHAT ANGLE DOES THE LINE 1-2 MAKE WITH THE PLANE? DRAW THE TRUE-SIZE OF THE PLANE AND COMPUTE ITS AREA.
METRIC SCALE: 1:1000

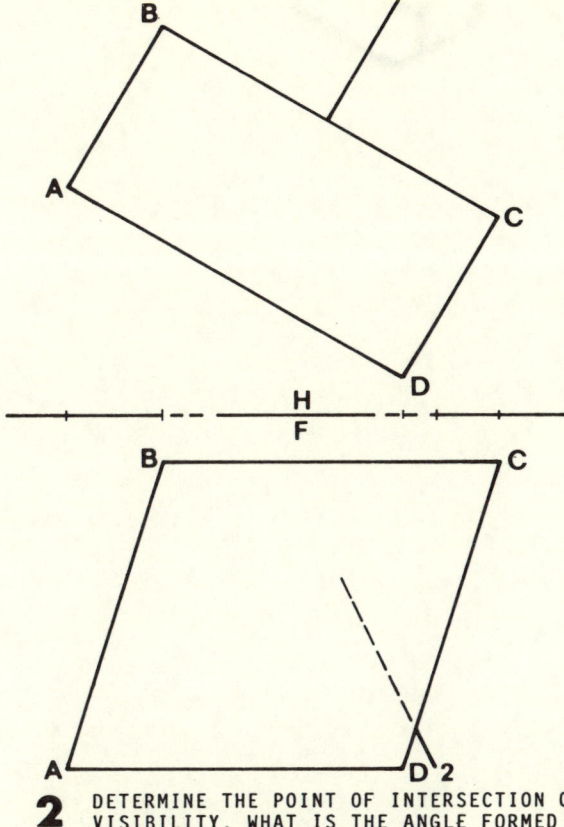

ANGLE =
AREA =

2 DETERMINE THE POINT OF INTERSECTION OF THE LINE 1-2 AND THE PLANE A-B-C. SHOW COMPLETE VISIBILITY. WHAT IS THE ANGLE FORMED BETWEEN THE LINE AND PLANE? DRAW THE TRUE-SIZE OF THE PLANE AND COMPUTE ITS AREA.
METRIC SCALE: 1:2000

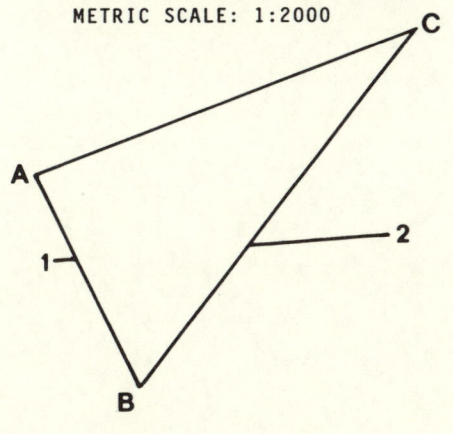

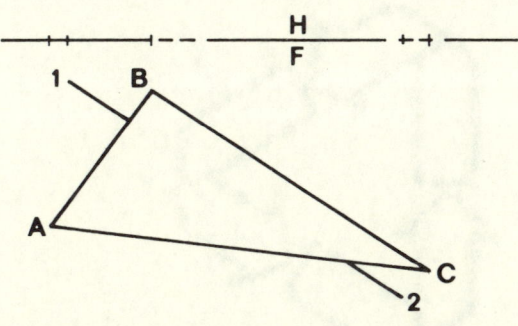

ANGLE =
AREA =

ANGLES BETWEEN LINES & PLANES | NAME | SECTION | DATE | GRADE | 94

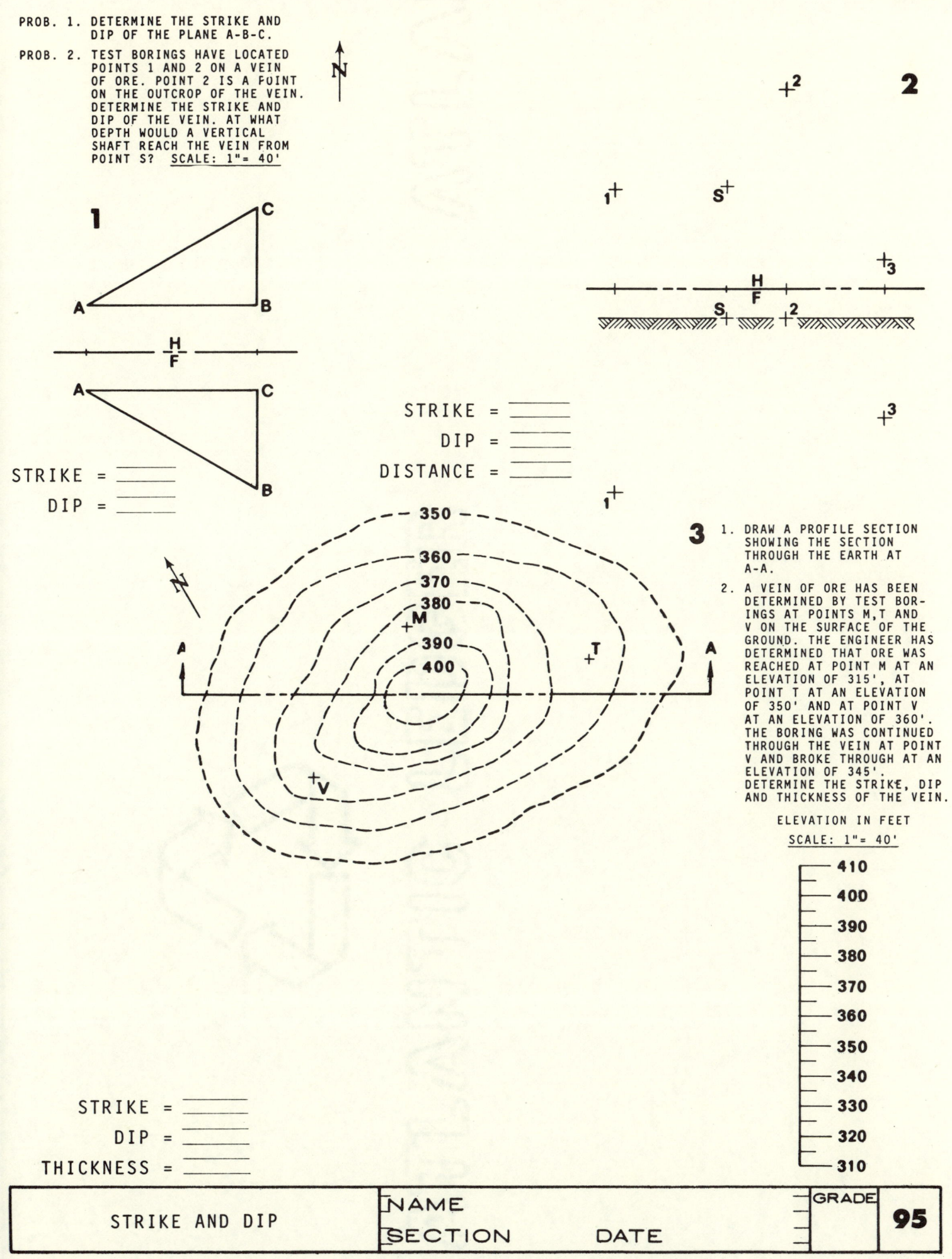

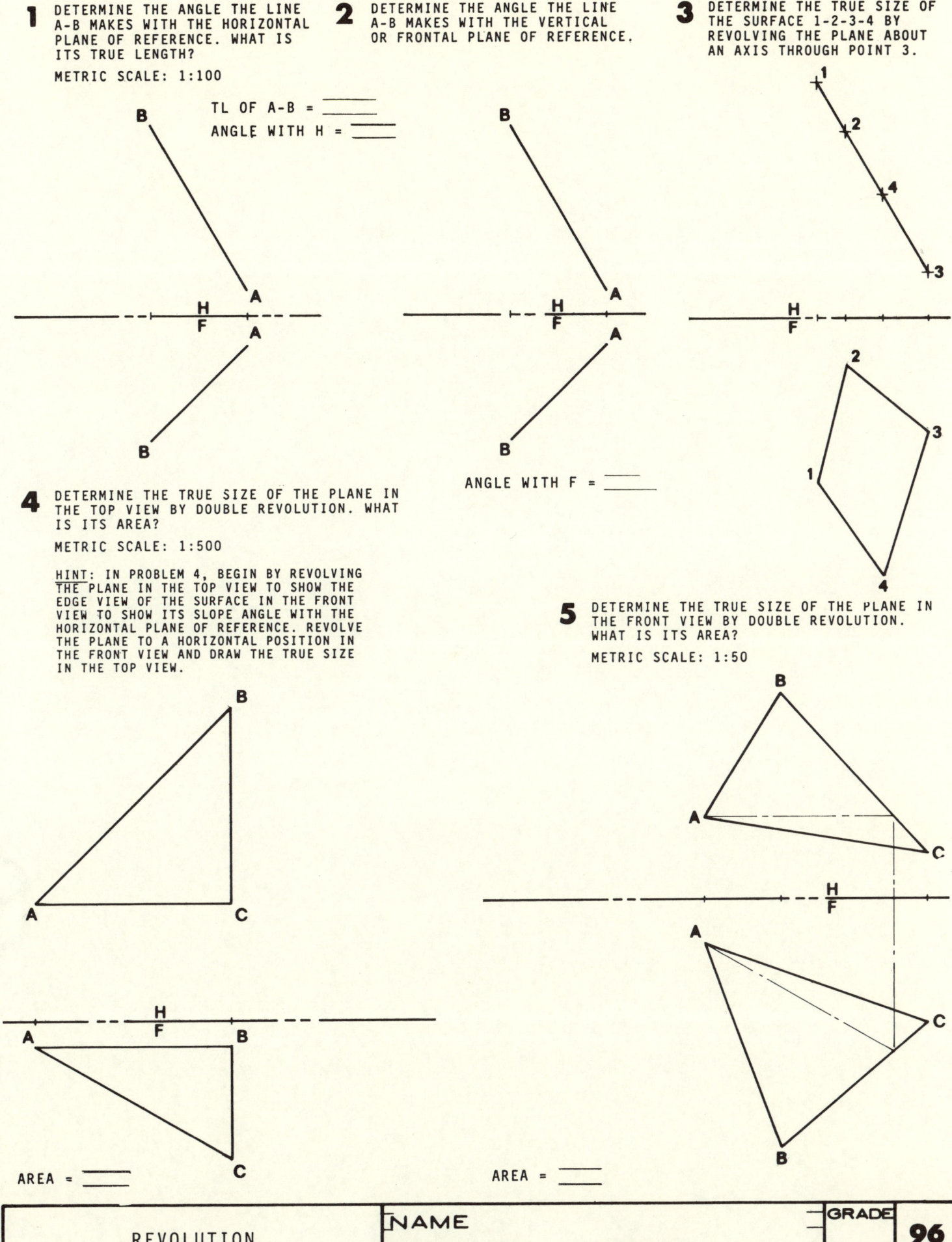

THE DRAWING BELOW IS A DESIGN FOR A MODEL ROCKET LAUNCHING PAD. IN ORDER TO BUILD THE DEVICE IT IS NECESSARY TO KNOW THE FOLLOWING:

1. THE ANGLES FORMED BETWEEN THE SIDES AND THE LAUNCHING PAD BASE.
2. THE TRUE-SIZE OF EACH SURFACE OF THE LAUNCHING PAD.

SPACE GEOMETRY PROBLEM

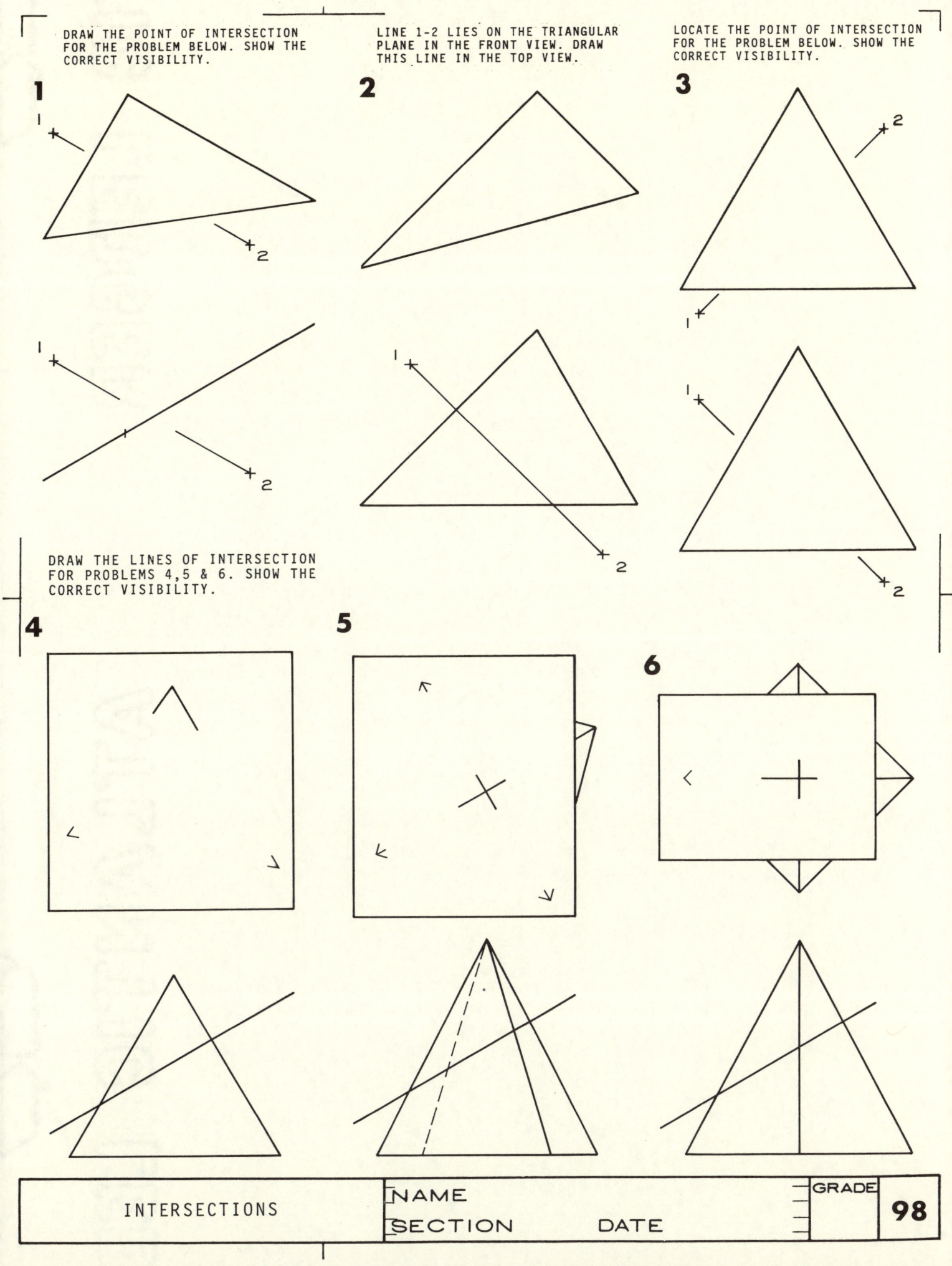

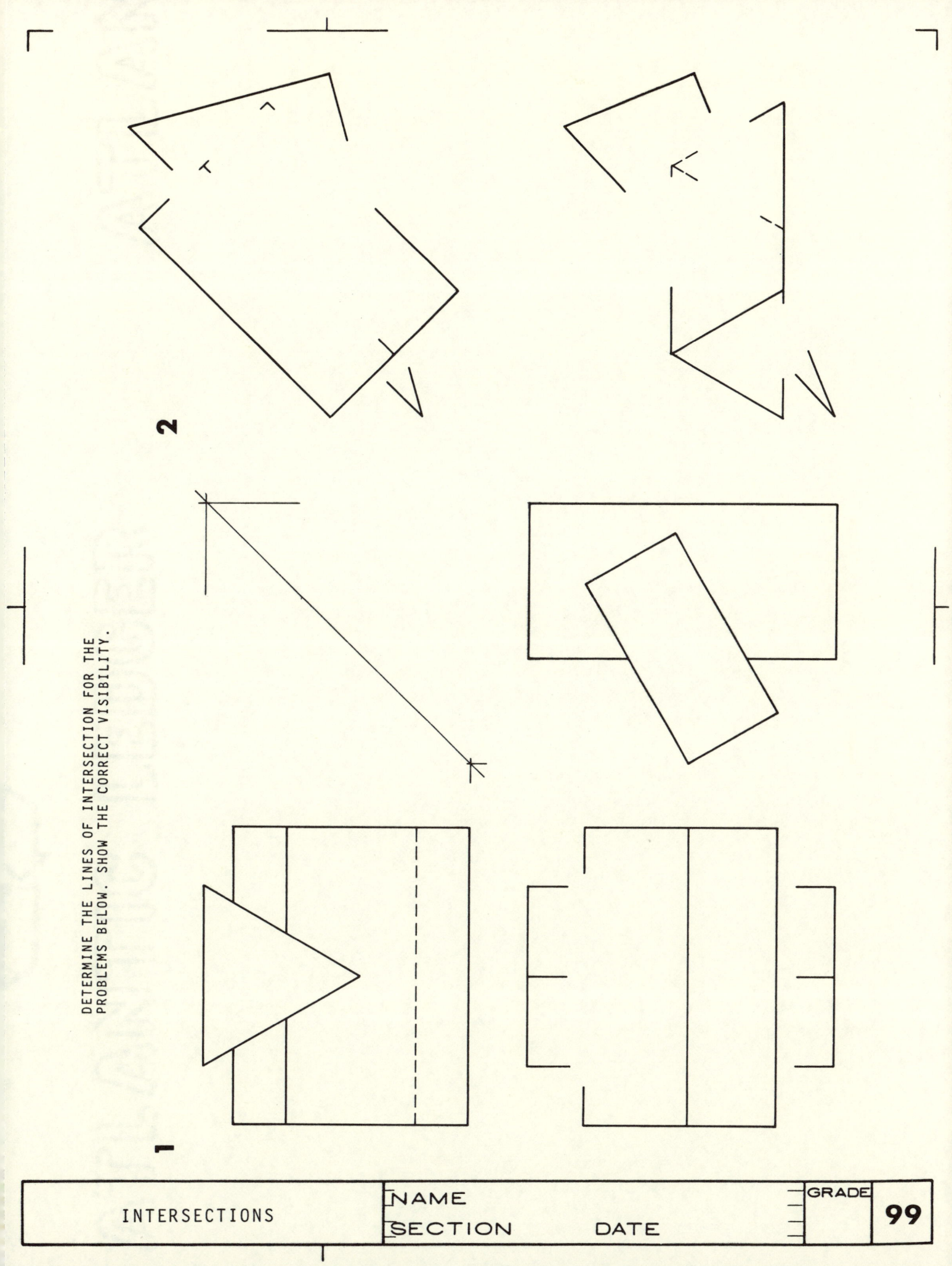

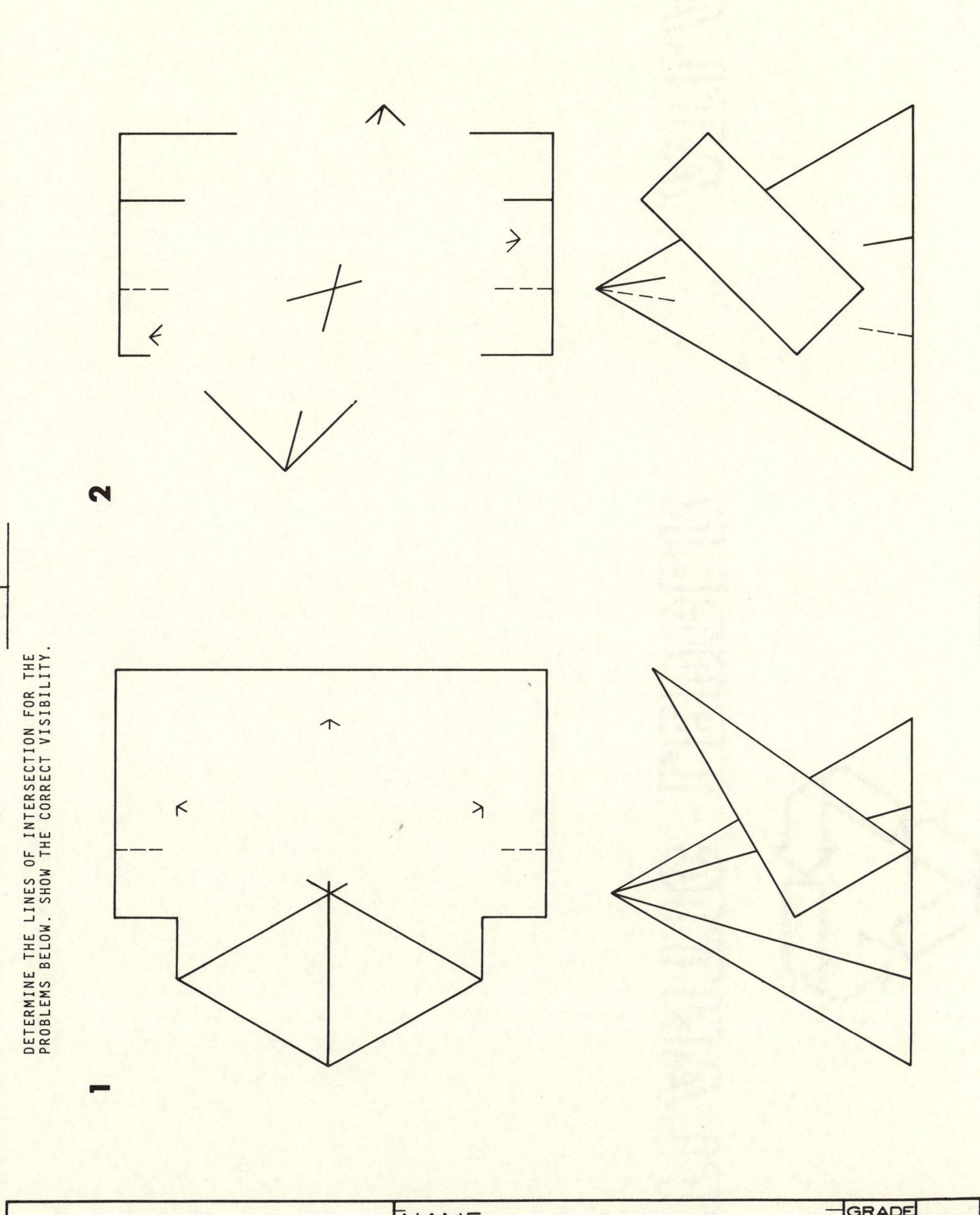

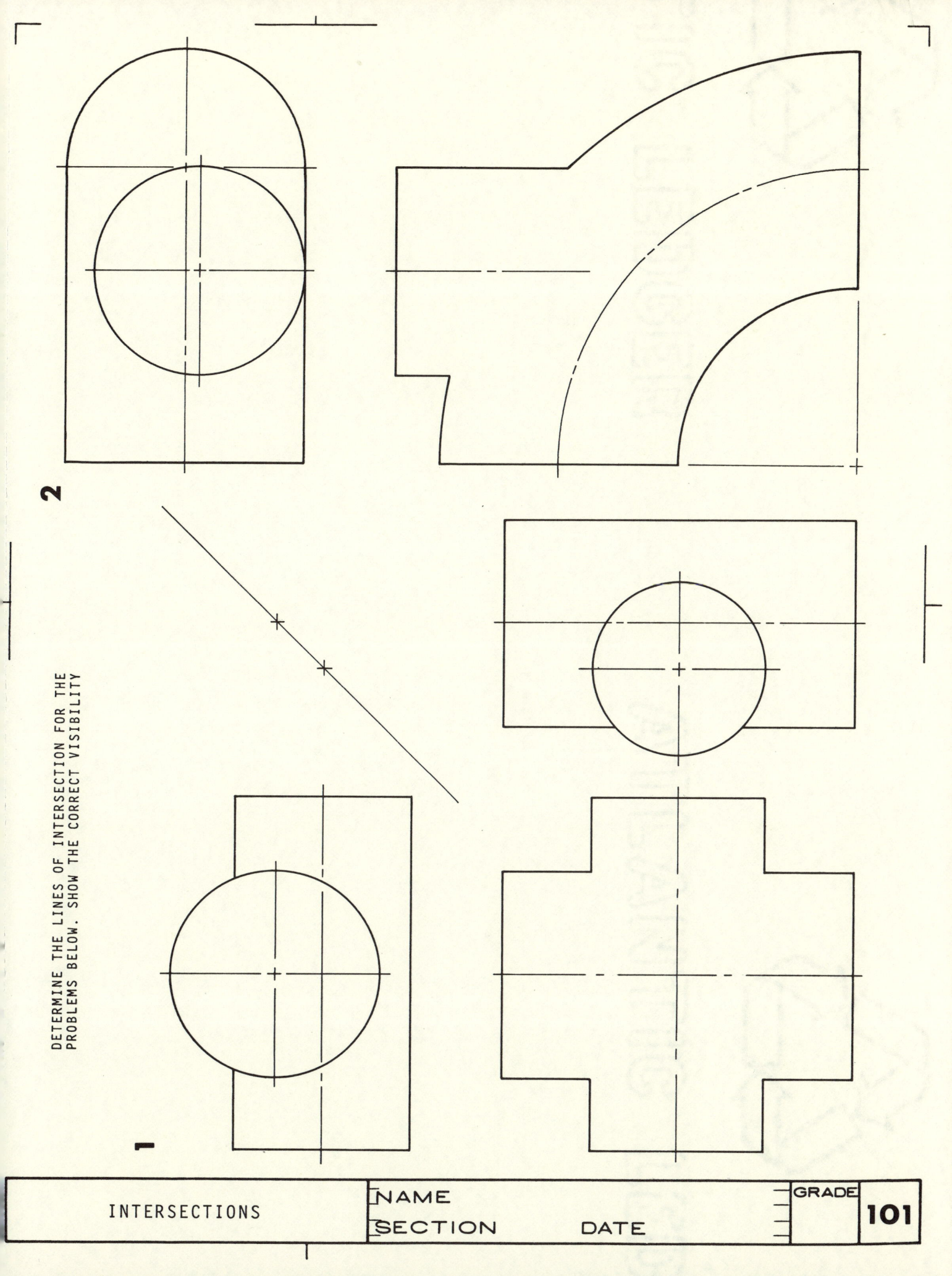

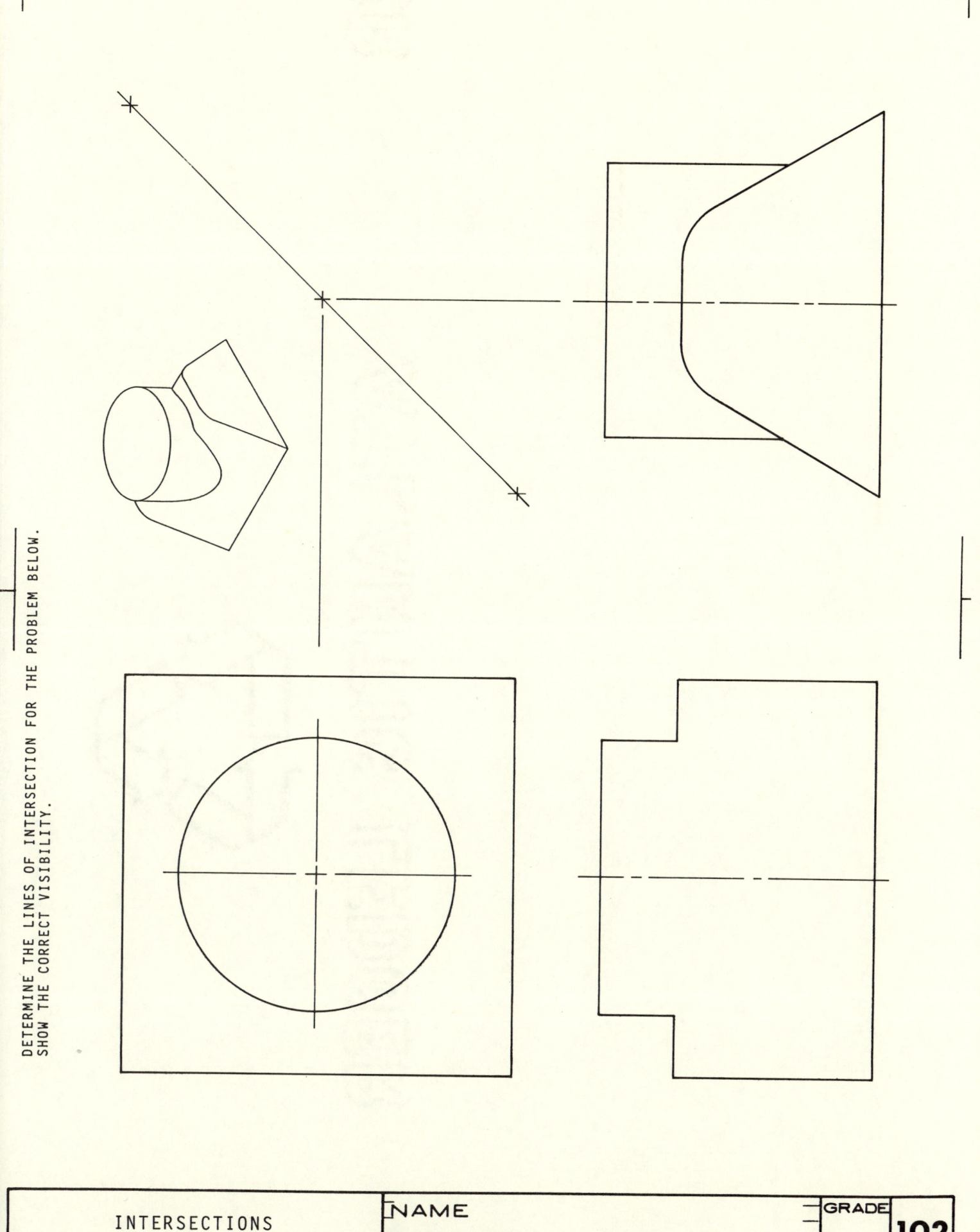

DRAW THE DEVELOPMENT OF THE IRREGULAR CONDUCTOR. BEGIN THE
PATTERN WITH THE LINE A-B AND DRAW WITH THE INSIDE SURFACE UP.

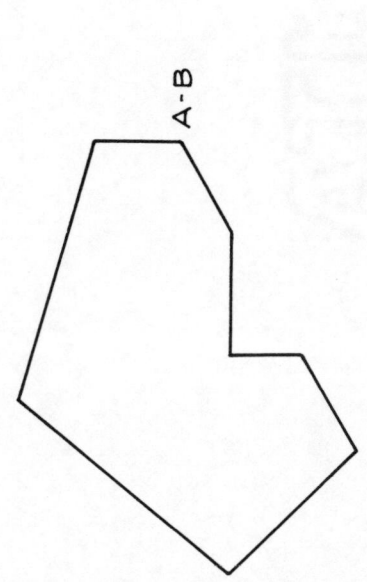

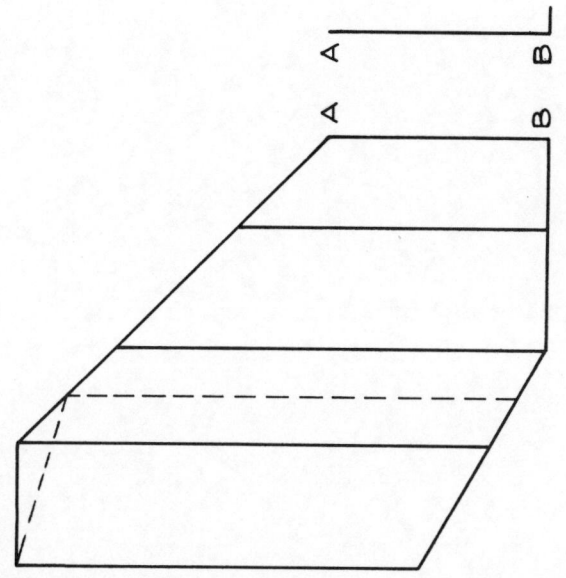

| DEVELOPMENTS | NAME | | GRADE | 103 |
| | SECTION | DATE | | |

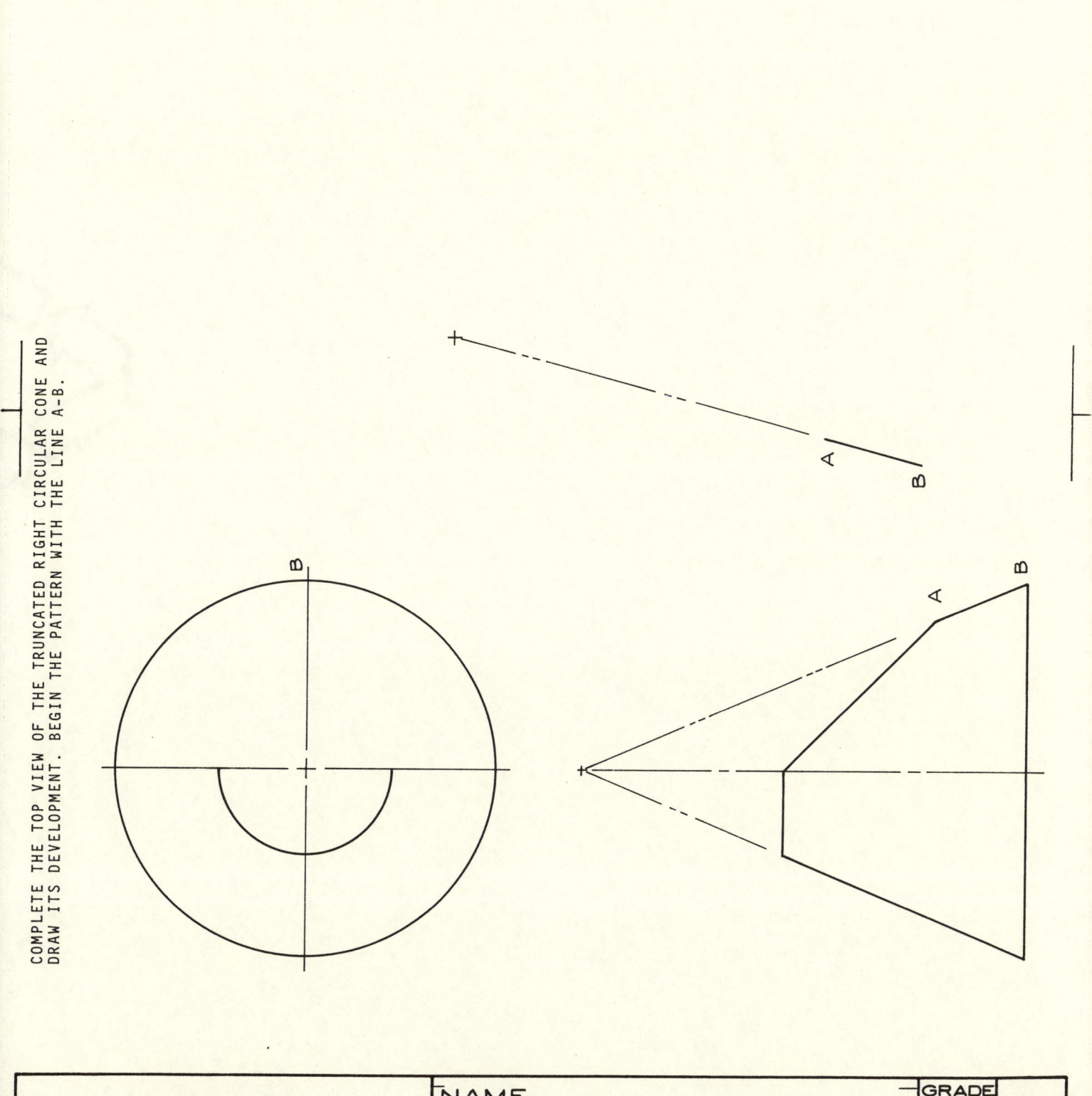

DEVELOPMENTS | NAME SECTION DATE | GRADE 104

COMPLETE THE TOP VIEW OF THE TRUNCATED HEXAGONAL PYRAMID AND
DRAW ITS DEVELOPMENT. BEGIN THE PATTERN WITH THE LINE A-B.

| DEVELOPMENTS | NAME SECTION DATE | GRADE 105 |

DRAW THE DEVELOPMENT OF THE LATERAL SURFACE OF THE TRANSITION PIECE.
BEGIN THE PATTERN WITH THE LINE A-B AND DRAW WITH THE INSIDE SURFACE UP.

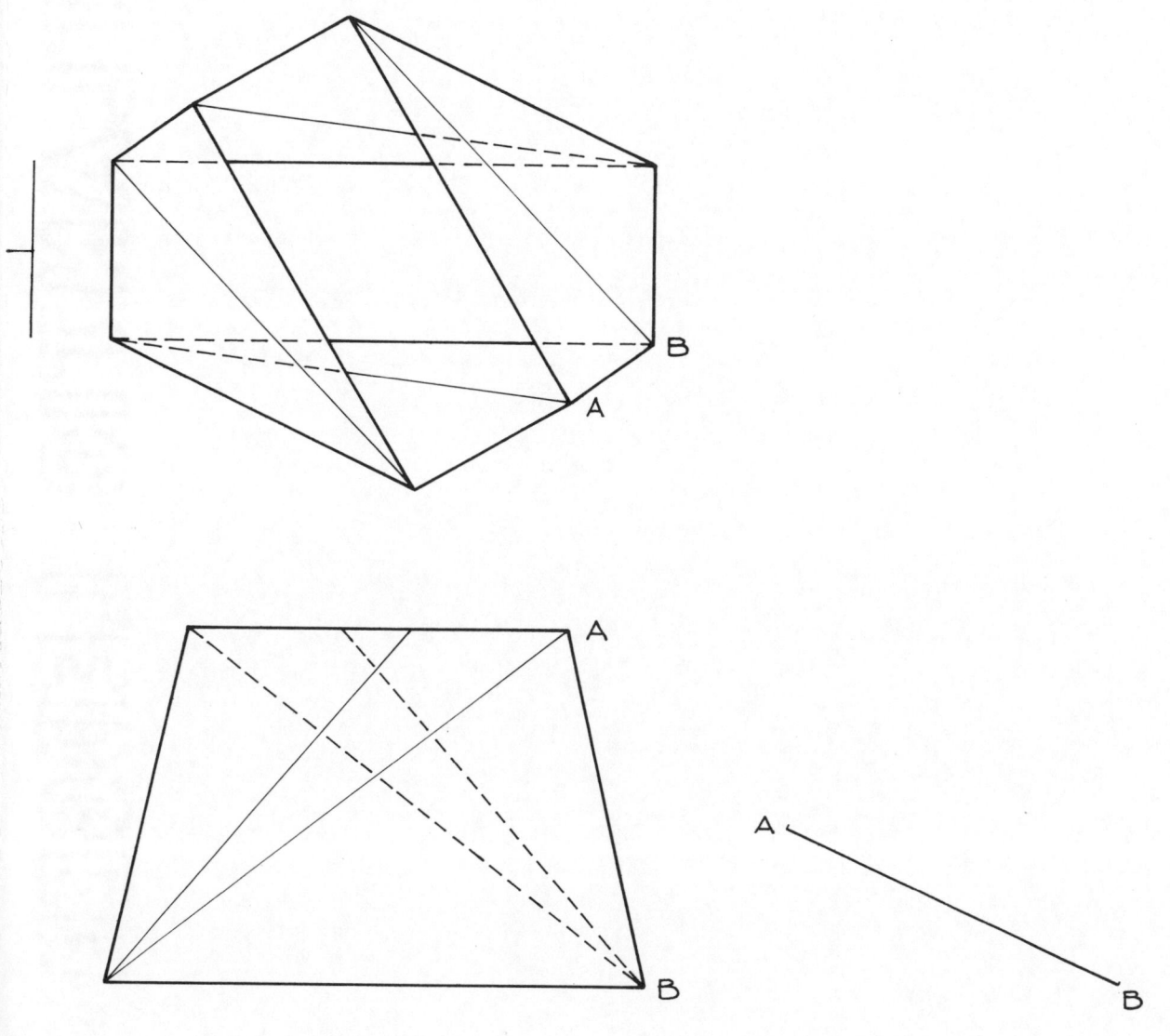

DEVELOPMENTS	NAME	GRADE 106
	SECTION DATE	

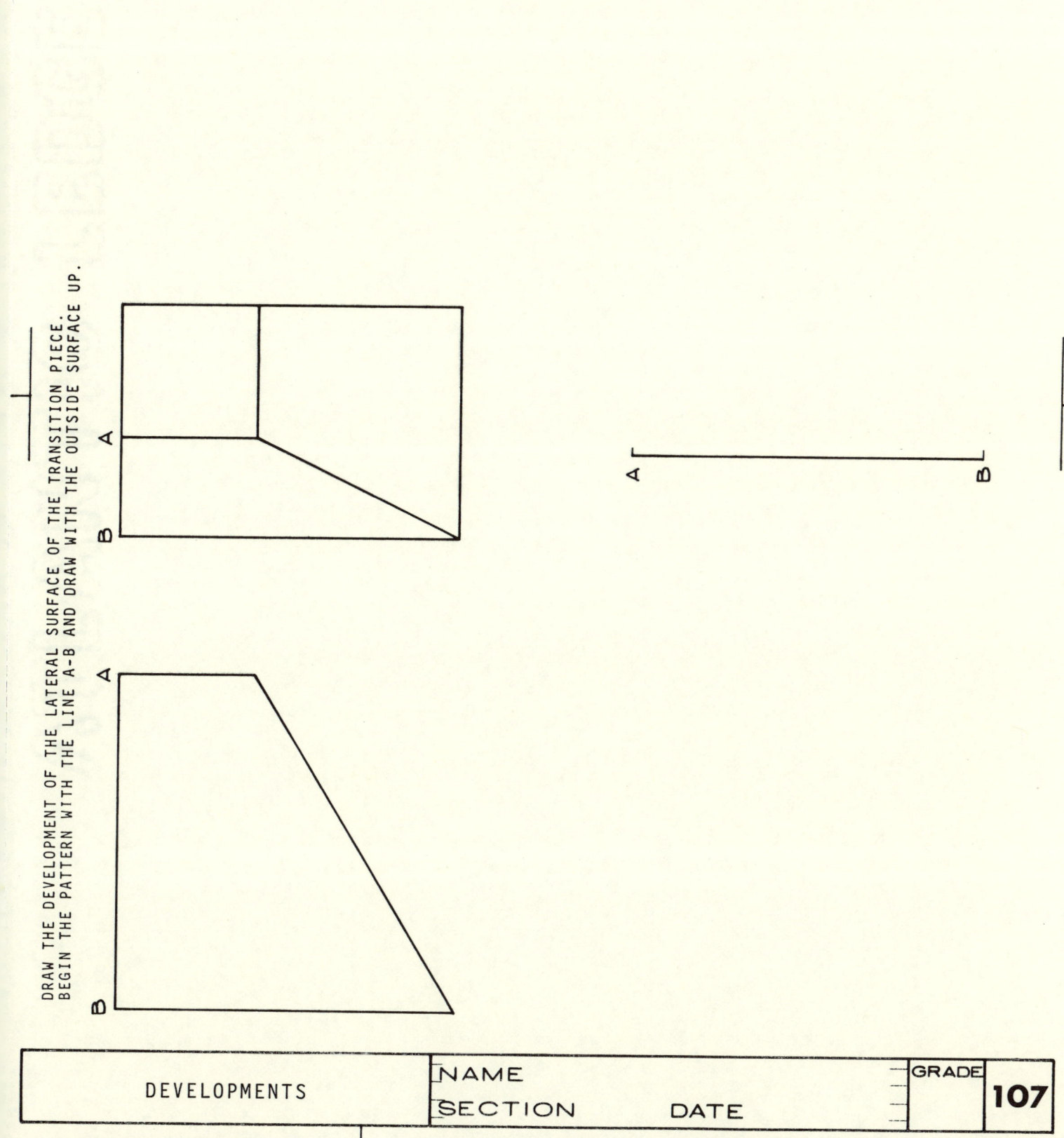

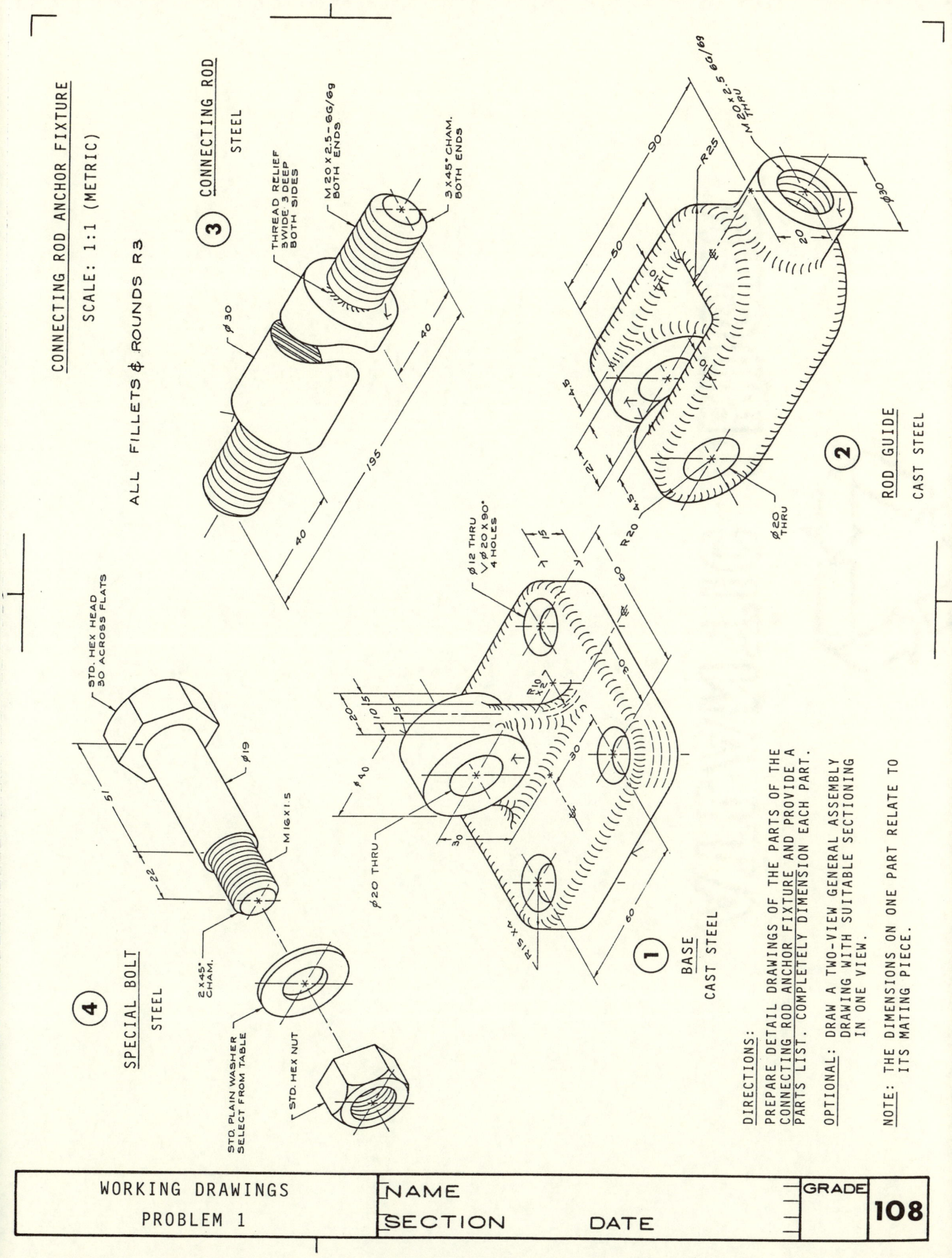

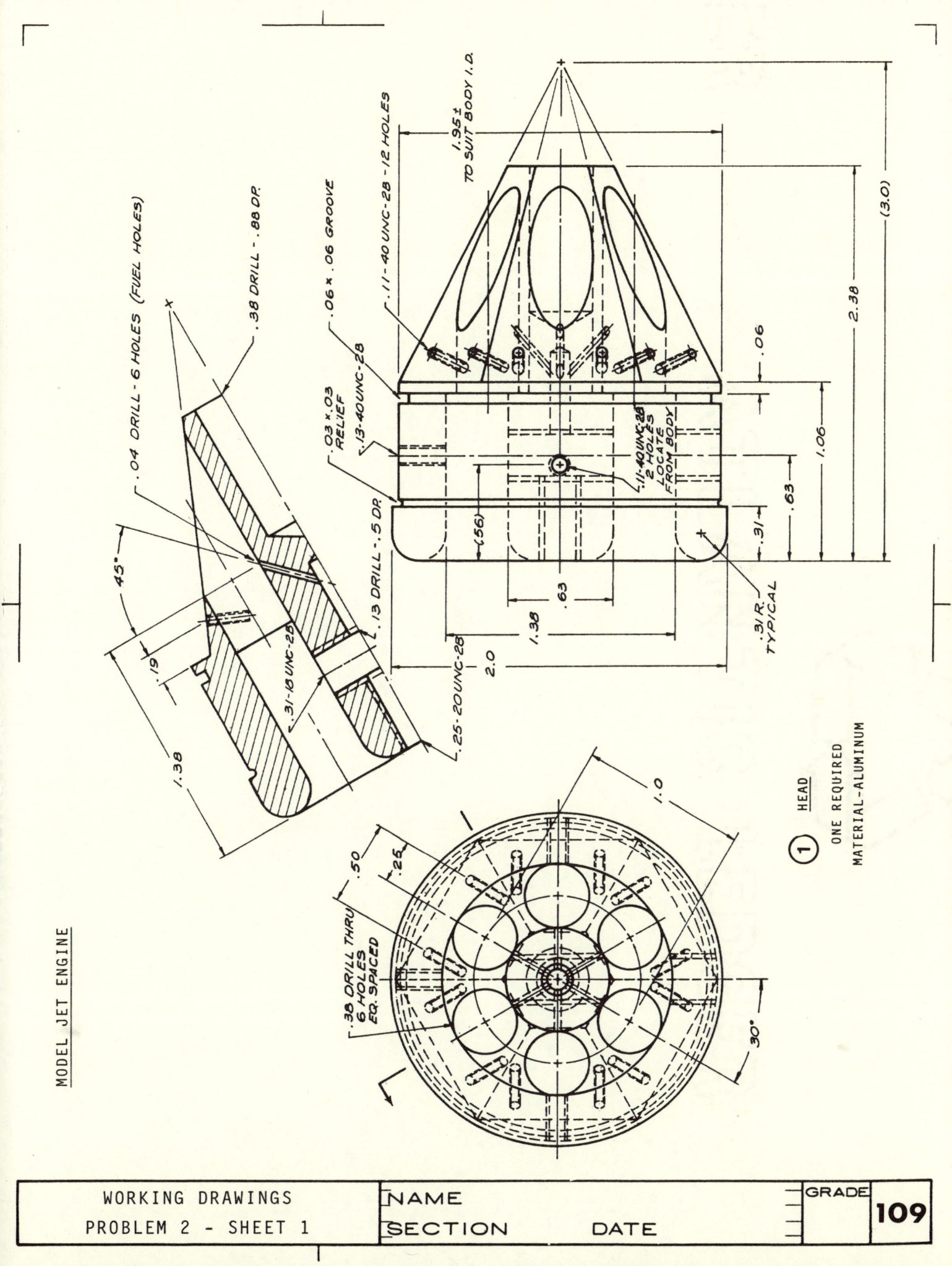

MODEL JET ENGINE

② BODY
ONE REQUIRED
MATERIAL – STAINLESS STEEL 26-23 GAGE

⑧ SPARK PLUG SEAT
ONE REQUIRED
MATERIAL – BRASS

⑦ PRESSURE LINE CONNECTION
ONE REQUIRED
MATERIAL – BRASS

| WORKING DRAWINGS PROBLEM 2 – SHEET 2 | NAME SECTION DATE | GRADE 110 |

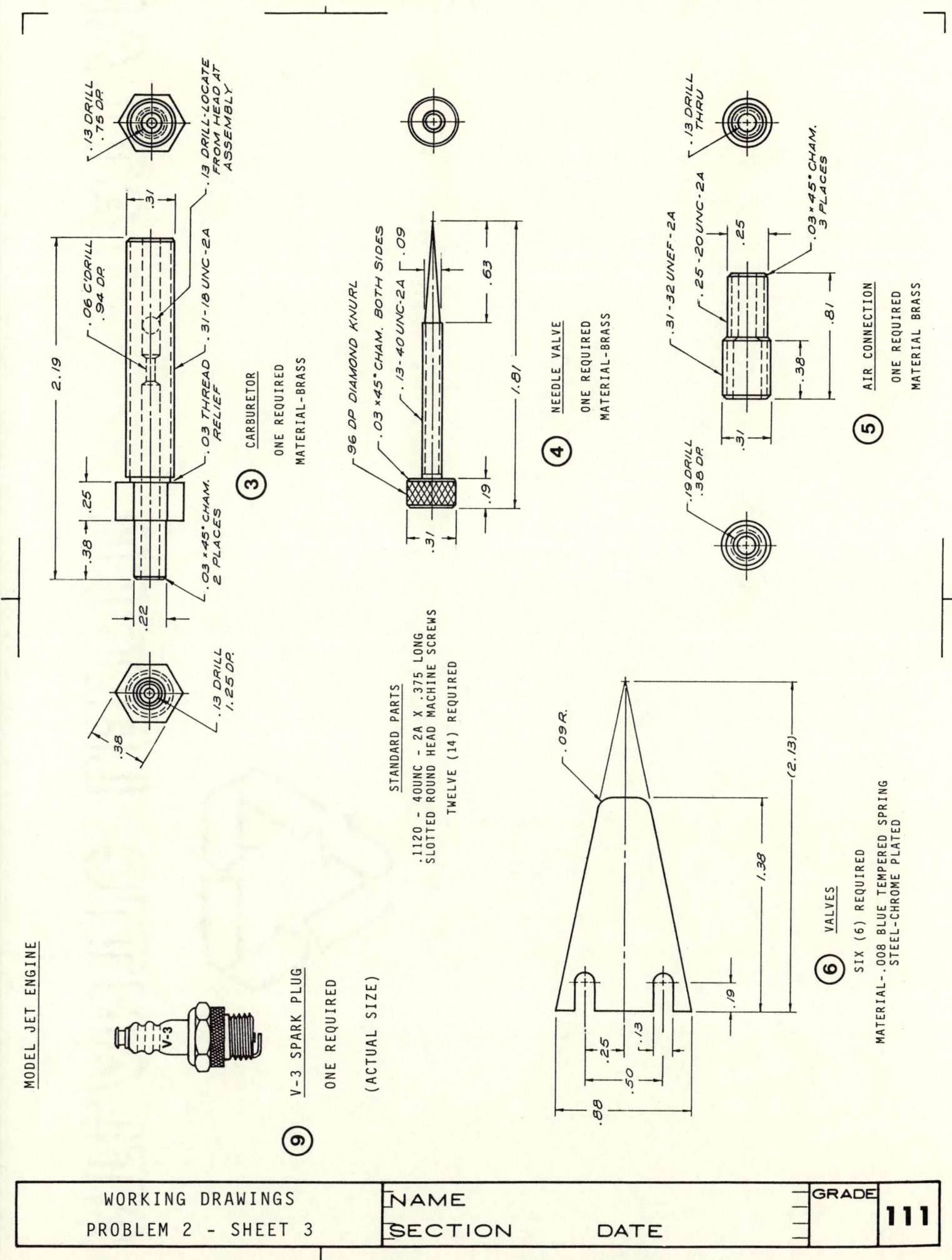

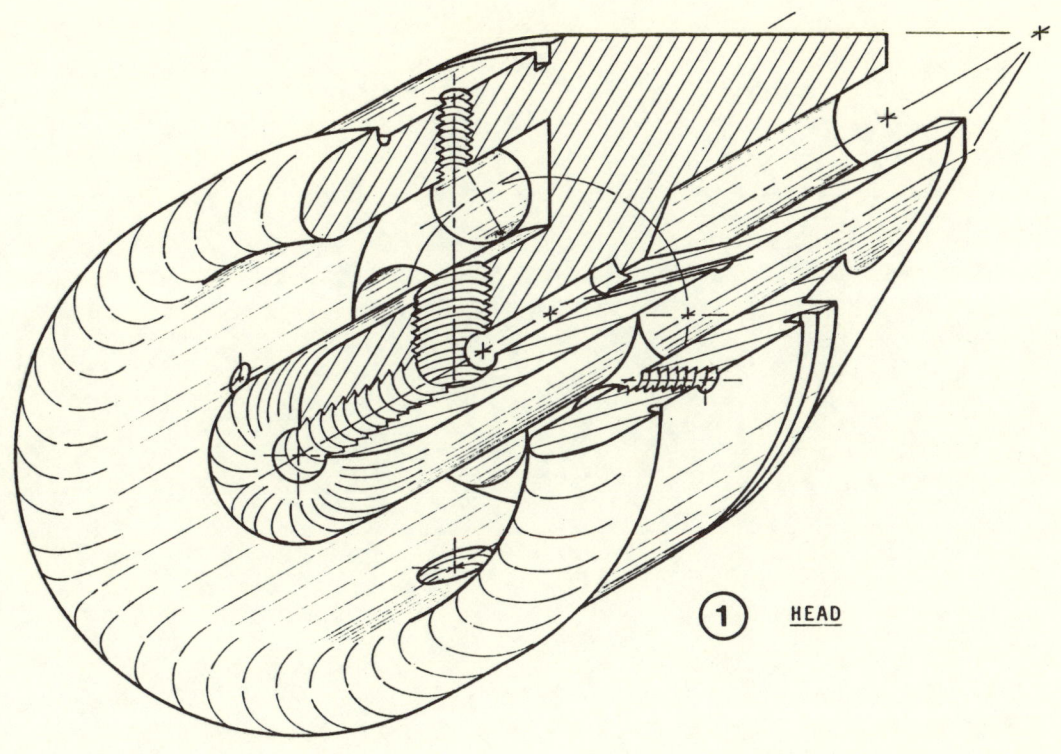

① HEAD

MODEL JET ENGINE

OPTIONS FOR DESIGN SOLUTION

YOUR INSTRUCTOR MAY ASSIGN THE FOLLOWING:

1. REDRAW ONE OR MORE OF THE PARTS (SCALE OPTIONAL) AND CONVERT THE DECIMAL INCHES TO MILLIMETERS.

2. DRAW A VIEW TO SHOW THE TRUE-SIZE OF THE VALVE SEAT.

3. DRAW A DOUBLE SIZE ORTHOGRAPHIC VIEW OF THE HEAD IN SECTION.

4. DRAW AN ASSEMBLY DRAWING OF ALL OF THE PARTS IN SECTION (SCALE OPTIONAL) AND PROVIDE A PARTS LIST.

5. DRAW AN ISOMETRIC DRAWING OF THE HEAD IN SECTION.

OVER LAY THE CIRCLE CHART BELOW WITH A CLEAR SHEET OF VELLUM AND DRAW A CIRCLE GRAPH TO REPRESENT THE PERCENTAGES OF PRODUCTS OBTAINED DURING THE REFINING OF PETROLEUM. REFER TO THE TABLE OF PERCENTS TO THE RIGHT. LABEL AND CROSSHATCH EACH SECTOR OF THE GRAPH. ADD A SUITABLE TITLE FOR THE COMPLETED GRAPH.

PERCENT YIELD OF PRODUCTS	ANGLE
43% - GASOLINE	
20% - HEAVY FUEL OIL	
19% - LIGHT FUEL OIL	
9% - ASPHALT, WAX, ETC.	
6% - KEROSENE	
3% - LUBRICATING OIL	

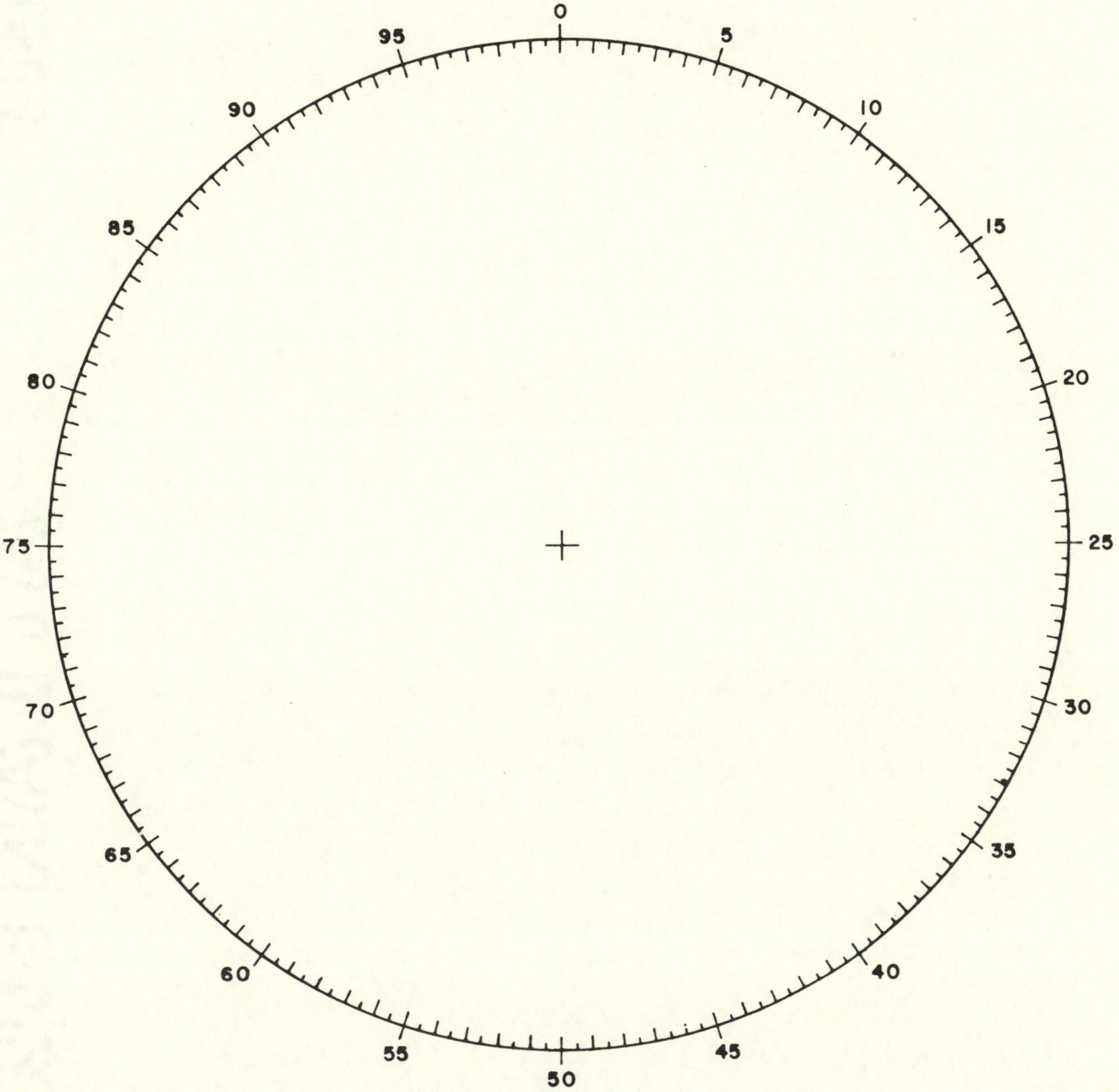

GRAPHS

NAME

SECTION   DATE

GRADE

113

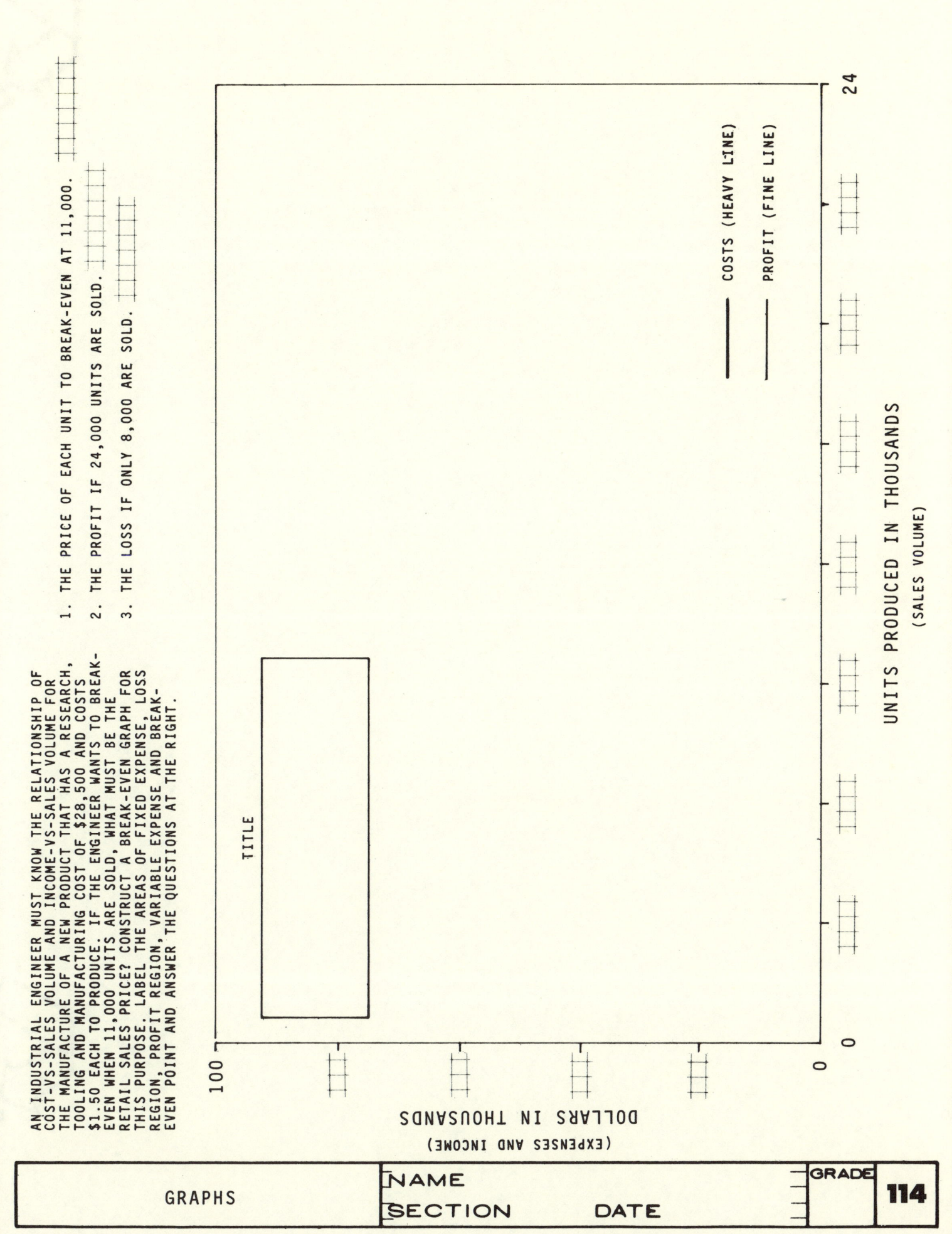

AN INDUSTRIAL ENGINEER MUST KNOW THE RELATIONSHIP OF COST-VS-SALES VOLUME AND INCOME-VS-SALES VOLUME FOR THE MANUFACTURE OF A NEW PRODUCT THAT HAS A RESEARCH, TOOLING AND MANUFACTURING COST OF $28,500 AND COSTS $1.50 EACH TO PRODUCE. IF THE ENGINEER WANTS TO BREAK-EVEN WHEN 11,000 UNITS ARE SOLD, WHAT MUST BE THE RETAIL SALES PRICE? CONSTRUCT A BREAK-EVEN GRAPH FOR THIS PURPOSE. LABEL THE AREAS OF FIXED EXPENSE, LOSS REGION, PROFIT REGION, VARIABLE EXPENSE AND BREAK-EVEN POINT AND ANSWER THE QUESTIONS AT THE RIGHT.

1. THE PRICE OF EACH UNIT TO BREAK-EVEN AT 11,000.
2. THE PROFIT IF 24,000 UNITS ARE SOLD.
3. THE LOSS IF ONLY 8,000 ARE SOLD.

GRAPHS — 114

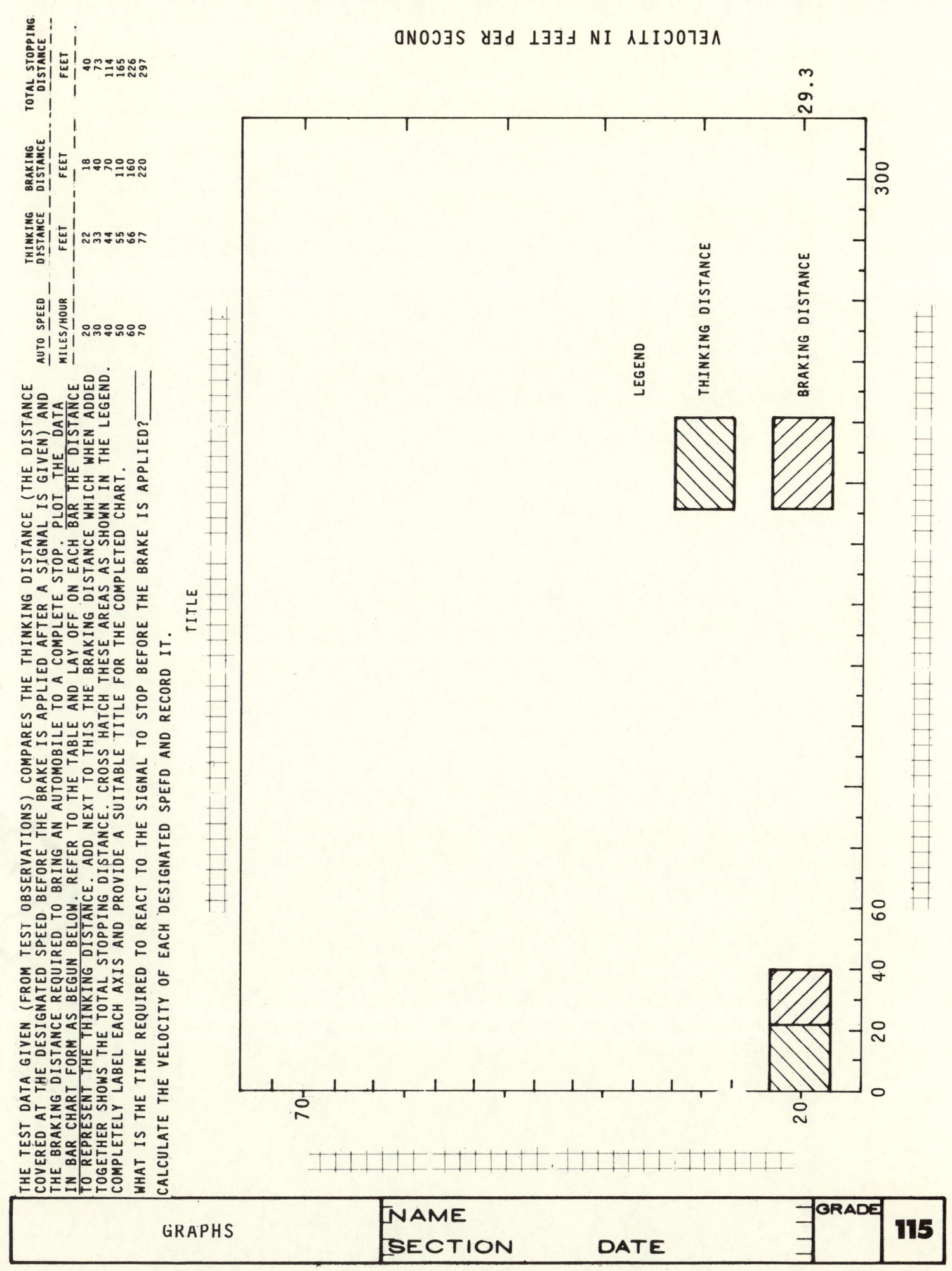

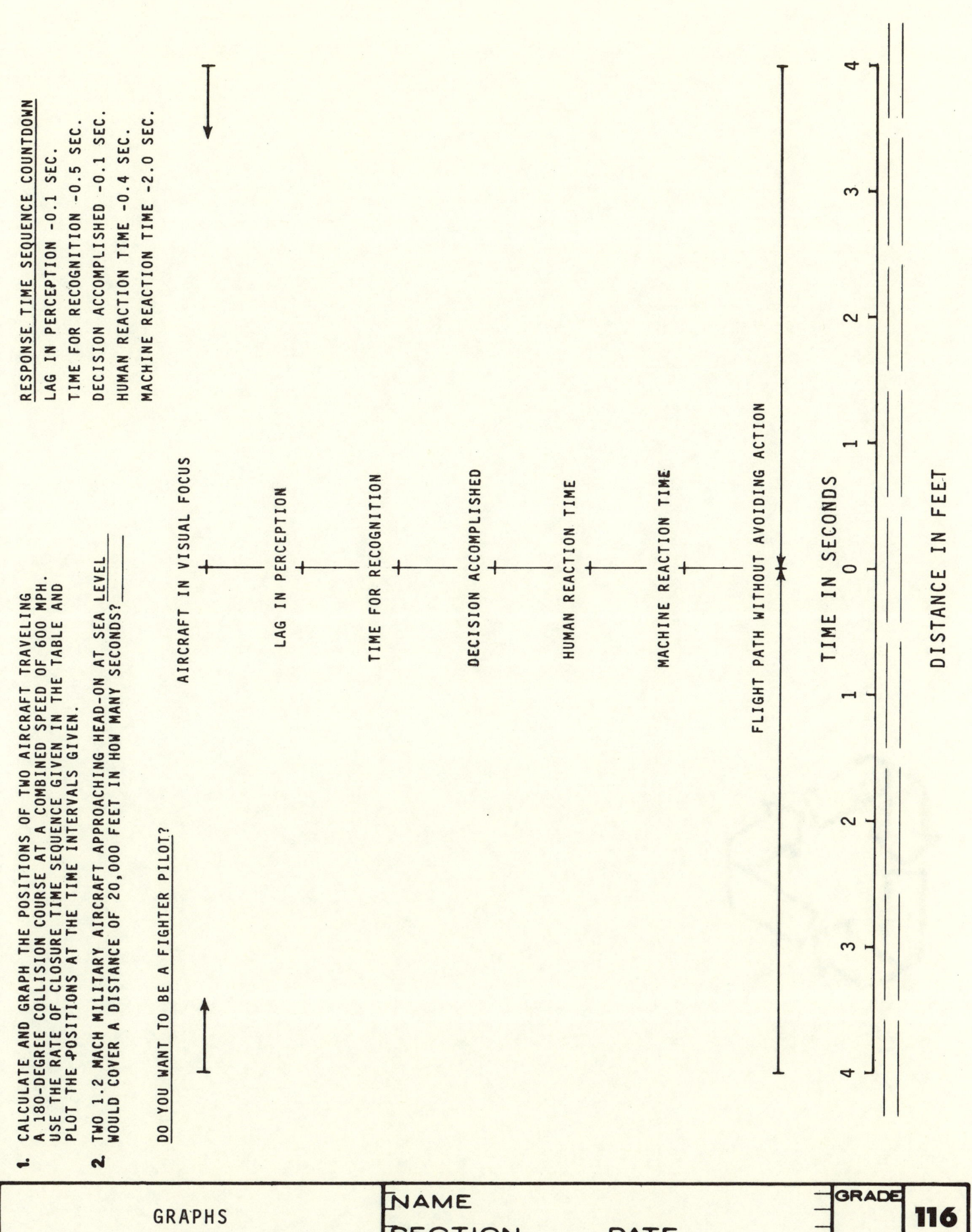

THE GIVEN CURVE WAS PLOTTED FROM EXPERIMENTAL DATA BASED ON OBSERVATIONS OF A COOLING BODY. DRAW THE DERIVATIVE CURVE TO DETERMINE THE RATE OF COOLING AT ANY INSTANT.
EXAMINE THE DATA TO ANSWER THE FOLLOWING:
1. WHAT IS THE COOLING RATE AT THE FIRST 5 SECONDS? _____
2. WHAT IS THE COOLING RATE AT THE FIRST 20 SECONDS? _____

GRAPHICAL DIFFERENTIATION

THE GIVEN CURVE IS THE PLOT OF AN OBJECT BEING MOVED BY A FORCE OF VARYING MAGNITUDE.
CONSTRUCT THE INTEGRAL CURVE WHICH WILL SHOW THE TOTAL WORK DONE IN MOVING THE OBJECT
AND THE AMOUNT OF WORK DONE ON THE OBJECT OVER ANY DESIRED DISTANCE.
EXAMINE THE DATA TO ANSWER THE FOLLOWING:
1. WHAT IS THE TOTAL WORK DONE ON THE OBJECT? _____
2. WHAT IS THE WORK DONE ON THE OBJECT MOVING FROM 15 TO 30 FEET? _____

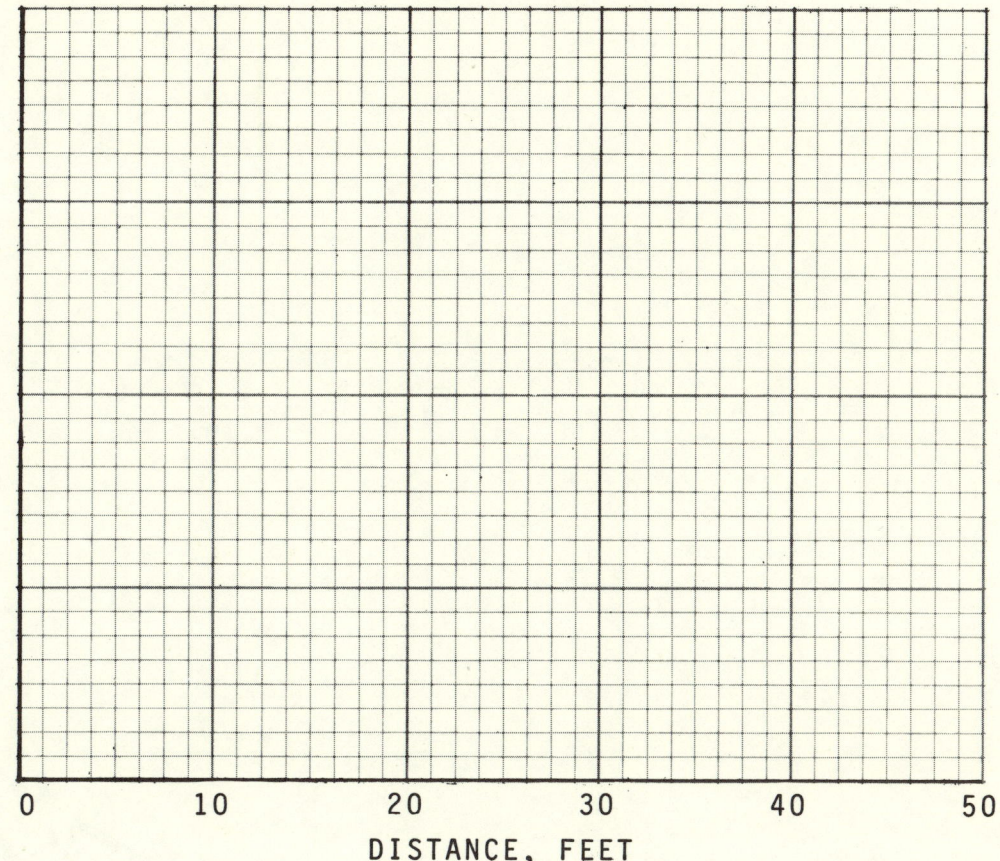

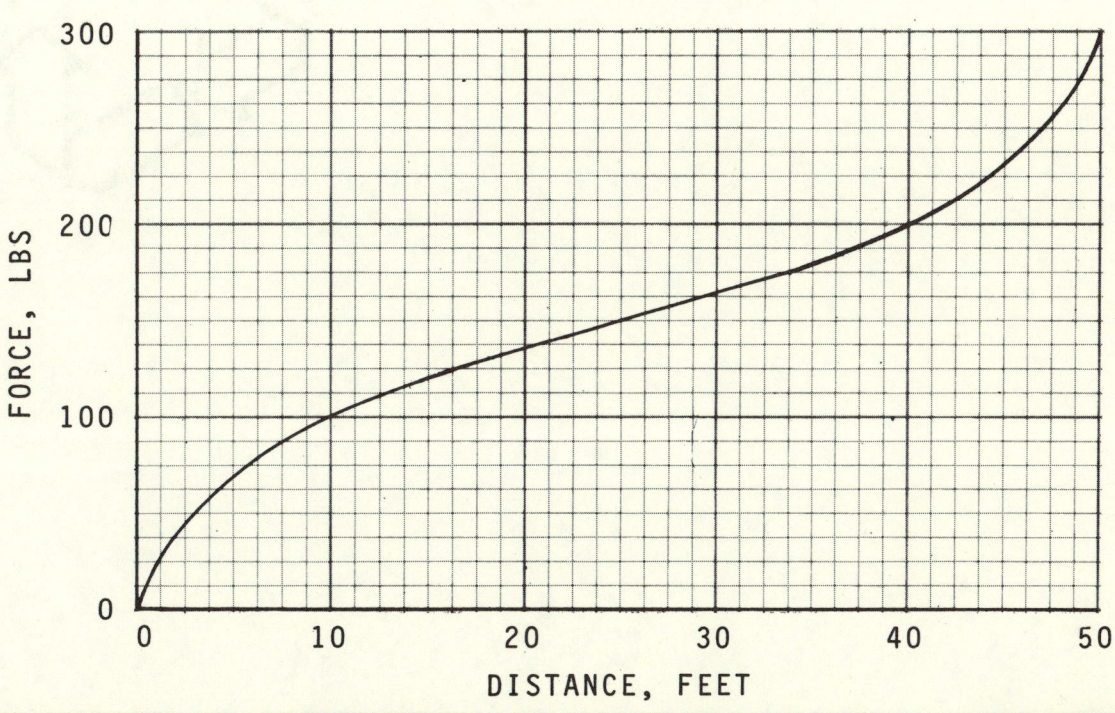

GRAPHICAL INTEGRATION

118

1 DETERMINE THE RESULTANTS OF THE TWO GIVEN VECTORS ACTING AT POINT O.
1. DETERMINE C IF C = A+B
2. DETERMINE D IF D = A-B

VECTOR SCALE: 1"= 100fps

2 GIVEN THE RESULTANT R. DRAW AND MEASURE THE FOLLOWING:
1. DRAW R AS TWO VECTORS SO THAT A+B = R
2. DRAW R AS TWO VECTORS SO THAT C+D = R

VECTOR SCALE: 1cm = 100N

A= _____   C= _____
B= _____   D= _____

3 DETERMINE THE FORCES IN THE CABLE AND BOOM WHEN THEY ARE ACTED UPON BY THE 100kg MASS HANGING FROM POINT O.

VECTOR SCALE: 1mm = 20N

BEGIN

4 DETERMINE THE FORCES IN THE STRUTS WHEN A REACTION OF F = 1500lb. IS APPLIED AS SHOWN.

VECTOR SCALE: 1"= 1000lb.

	FORCE	+ OR -
CABLE		
BOOM		

5 1. DETERMINE THE FORCES IN THE WEIGHTLESS STRUT AND THE CABLE.

VECTOR SCALE: 1"= 500lb.

2. IF THE FORCE IN THE CABLE CANNOT EXCEED 1300lb, DETERMINE THE DISTANCE ABOVE THE STRUT THAT THE CABLE MAY BE FASTENED. DRAW AND DIMENSION THIS NEW LOCATION. WHAT IS THE FORCE IN THE STRUT WHEN THE CABLE IS SO FASTENED?

SCALE: 1"= 6'

BEGIN

	FORCE	+ OR -
CABLE		
STRUT		

NEW LOCATION OF STRUT

FORCE = _____

VECTORS

1 THE FORCES IN THE SPACE DIAGRAM BELOW ARE DRAWN TO SCALE AS VECTORS. THE VECTORS ACT ON POINT O. DRAW THE VECTORS BY DOUBLING THE SPACE DIAGRAM WITH DIVIDERS. DETERMINE THE MAGNITUDE AND SENSE OF THE RESULTANT.

VECTOR SCALE: 1cm = 5N

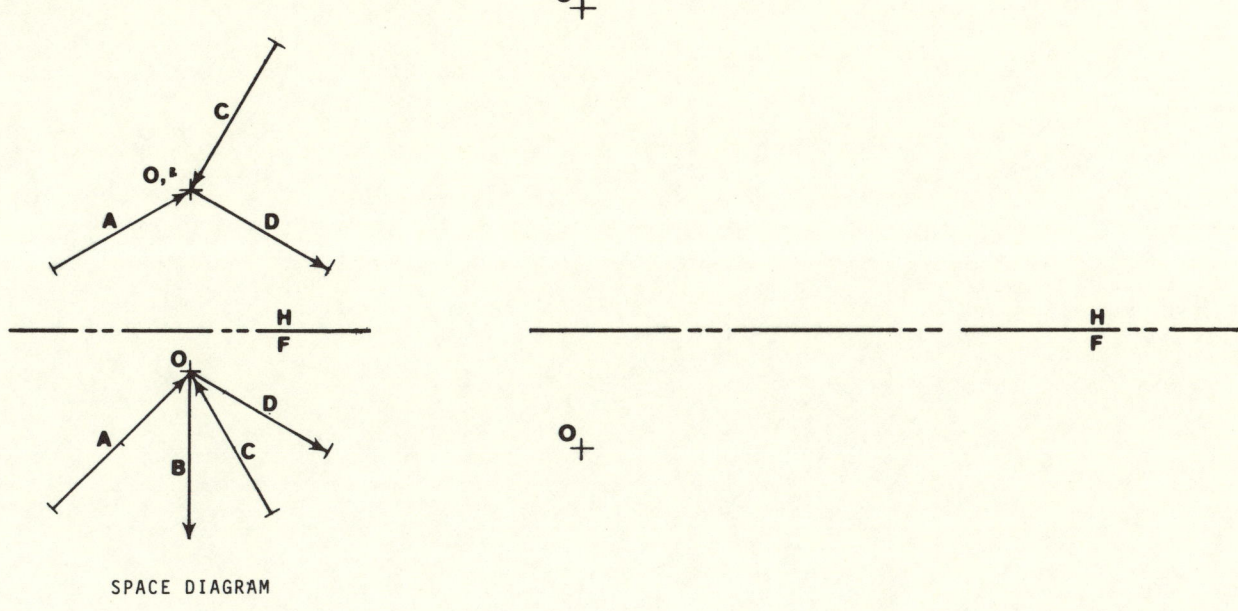

SPACE DIAGRAM

2 DETERMINE THE MAGNITUDE AND DIRECTION OF THE REACTIONS OF THE LANDING GEAR AT POINTS A AND B TO BALANCE THE LOAD OF 3000lb AS SHOWN.

VECTOR SCALE: 1" = 1000lb

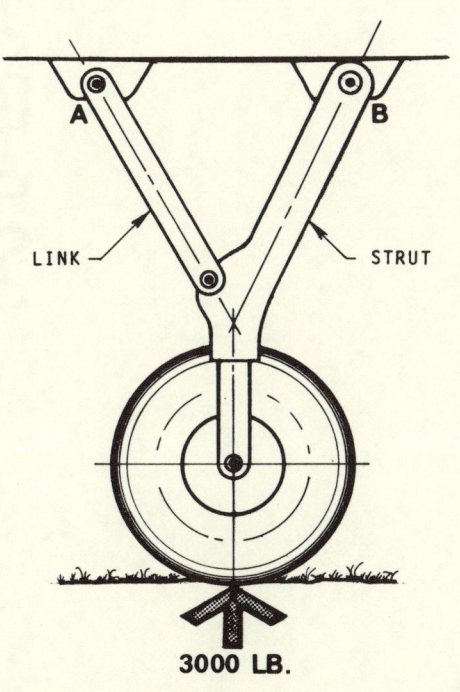

3000 LB.

BEGIN F

VECTORS

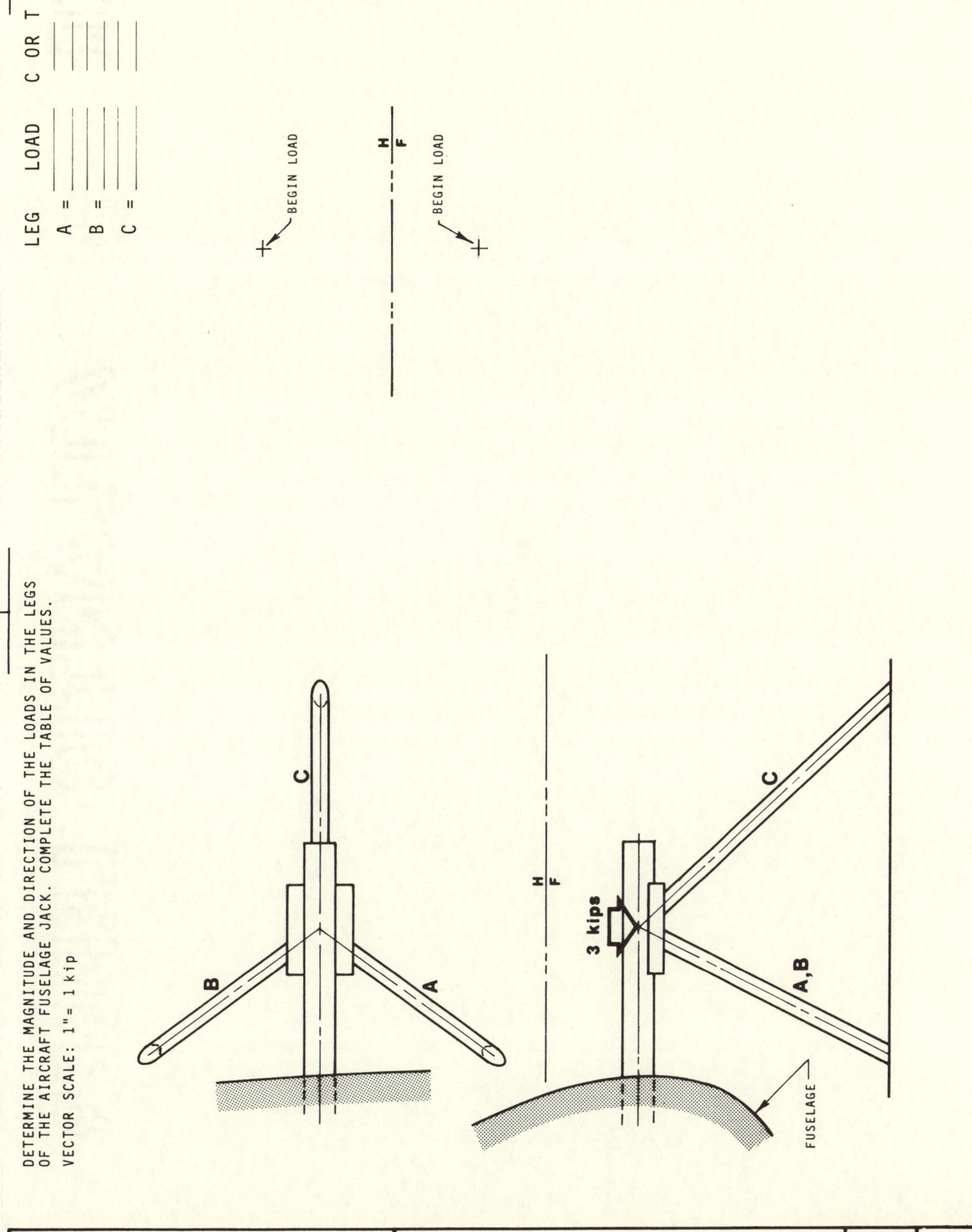

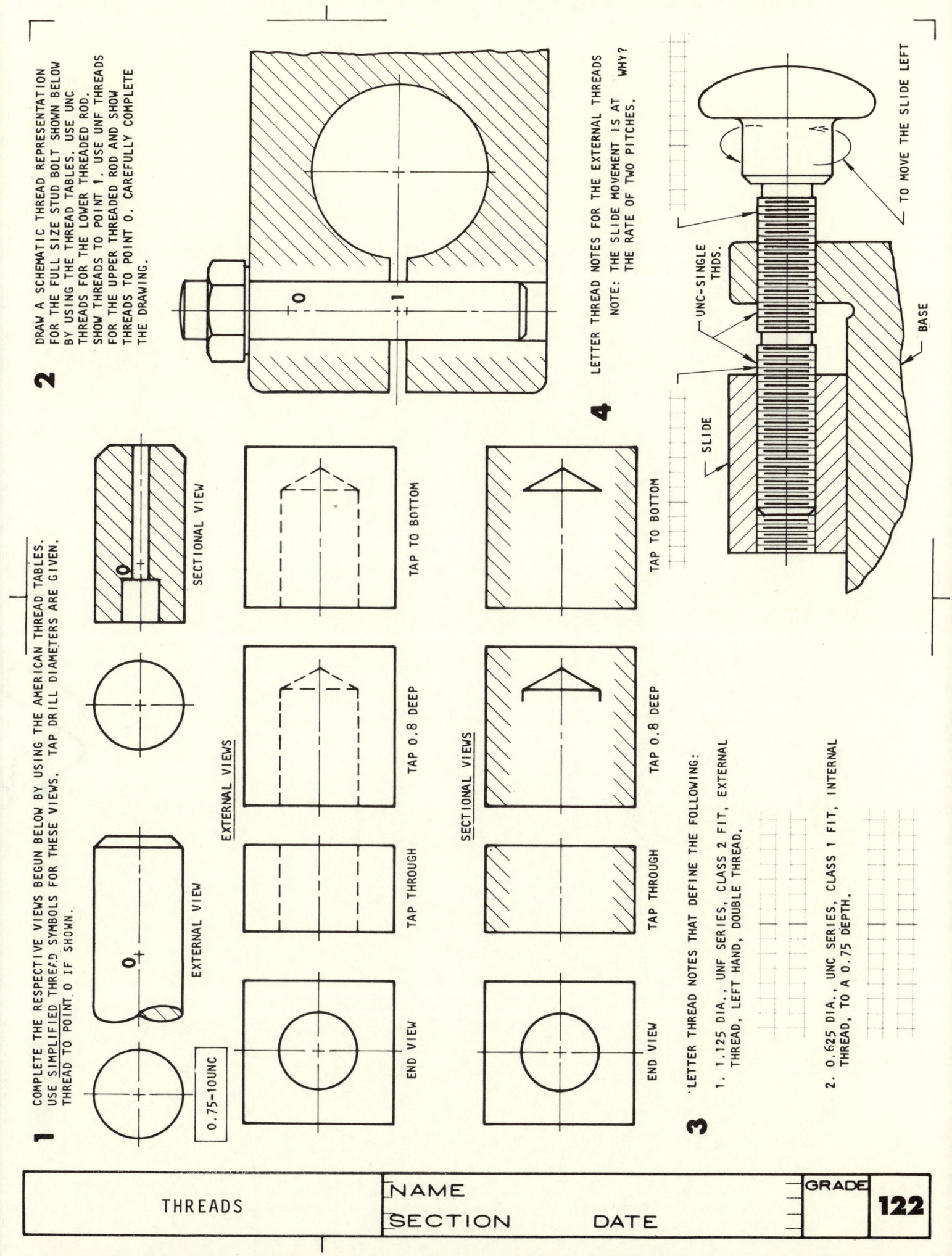

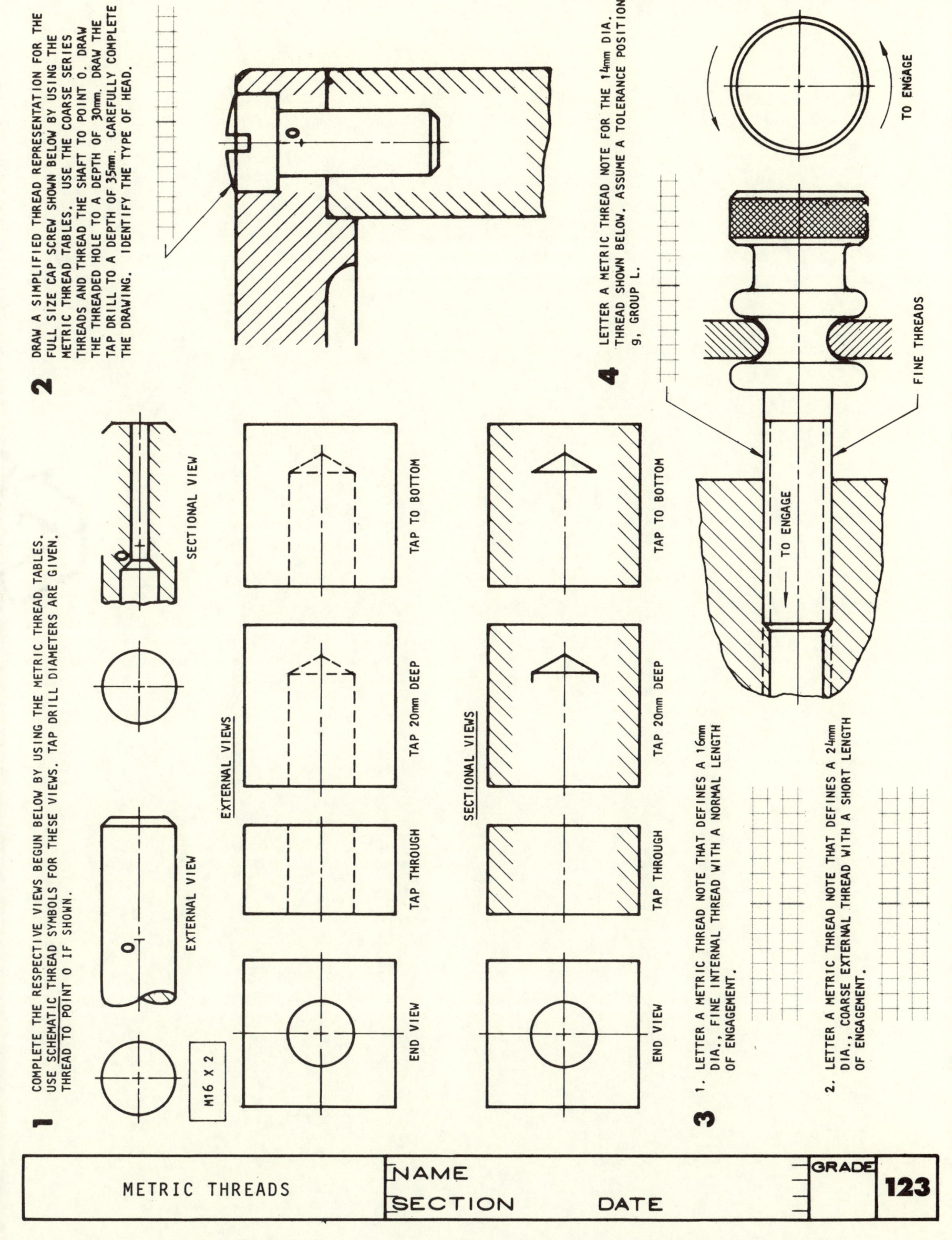

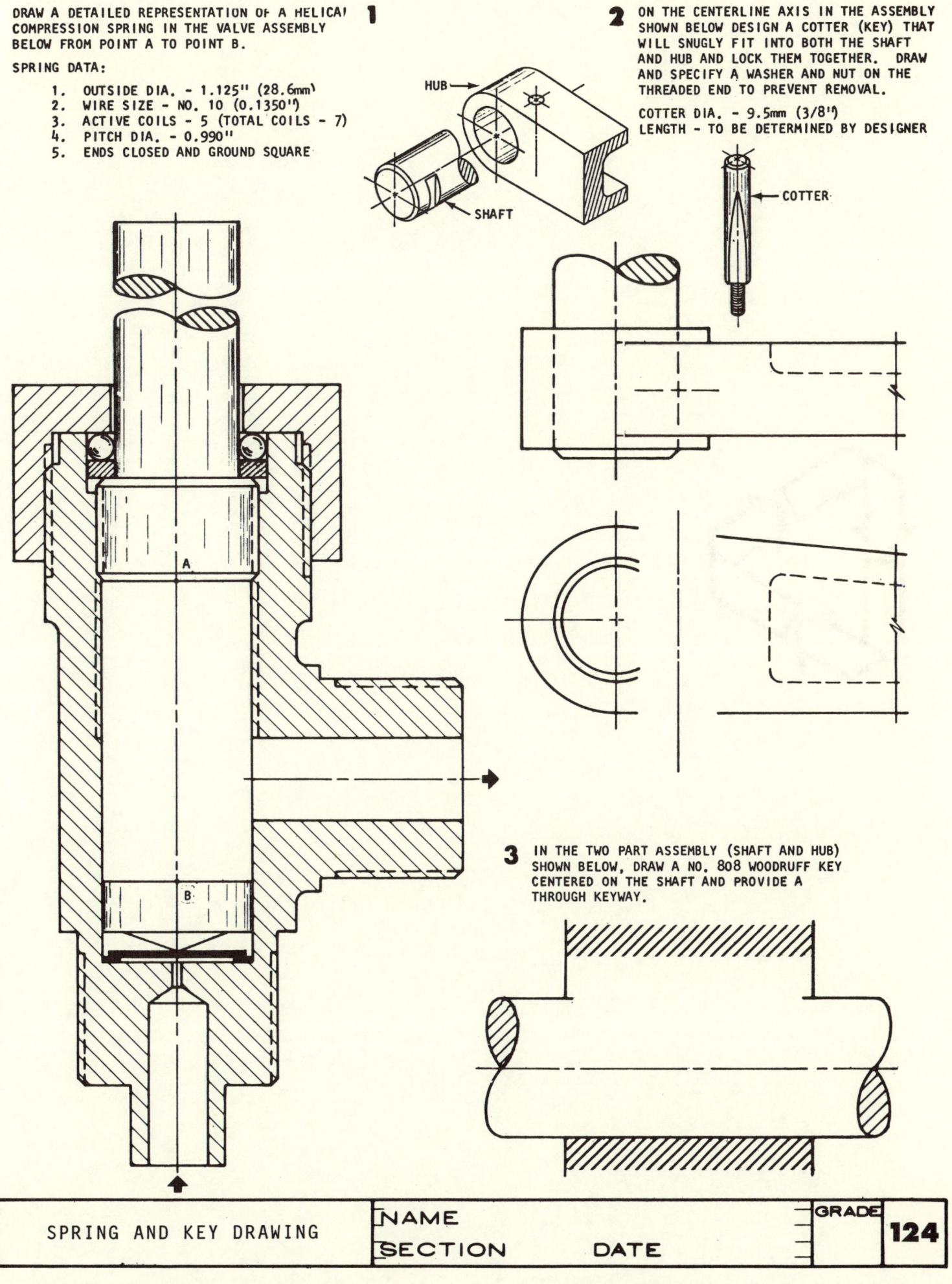

A WORM AND WORM GEAR COMBINATION TRANSMIT MOTION BETWEEN TWO SHAFTS WHICH ARE AT RIGHT ANGLES TO EACH OTHER BUT DO NOT INTERSECT. THE WORM IS THE DRIVER, AND EACH REVOLUTION OF THE WORM ROTATES THE WORM GEAR ONE TOOTH; OR IF A DOUBLE THREADED WORM, TWO TEETH.

DRAW A WORM AND WORM GEAR USING THE SPECIFICATIONS AS GIVEN BELOW. DRAW THE UPPER HALF OF THE WORM AS A SECTION. REFER TO THE TEXT APPENDIX AND DRAW A SQUARE KEY AND KEYWAY FOR THE WORM. DRAW THE WORM GEAR AND SHOW AT LEAST 8 TEETH. DRAW THE REMAINDER OF THE GEAR CONVENTIONALLY.

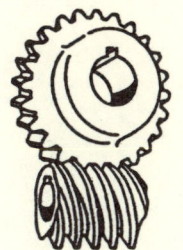

WORM

NO. OF THREADS	- 2
PITCH DIAMETER	- 2.000
AXIAL PITCH	- 0.524
WHOLE DEPTH	- 0.359
FACE LENGTH	- 3.000
OUTSIDE DIA.	- 2.320

SCALE: FULL SIZE

PRESSURE ANGLE 14½°

WORM GEAR

NO. OF TEETH	- 36
PITCH DIAMETER	- 6.000
CIRCULAR PITCH	- 0.524
WHOLE DEPTH	- 0.359
ADDENDUM	- 0.167
BASE CIRCLE	- 5.809

WHAT IS THE VELOCITY RATIO OF THE WORM GEAR? _____

WHAT IS THE CLEARANCE BETWEEN WORM AND THE WORM GEAR? _____

GEAR DRAWING

NAME
SECTION   DATE

GRADE **125**

DRAW THE DISPLACEMENT DIAGRAM AND
CAM PROFILE WITH THE FOLLOWING MOTION:

1. DWELL 30°
2. LIFT 2" IN 120° WITH HARMONIC MOTION
3. DWELL 60°
4. FALL 2" IN 120° WITH HARMONIC MOTION
5. DWELL 30°

THE CAM ROTATES COUNTER-CLOCKWISE.

SHOW THE POSITION OF THE ROLLER FOLLOWER
AT 30° INTERVALS ABOUT THE CAM.

SCALE: FULL SIZE

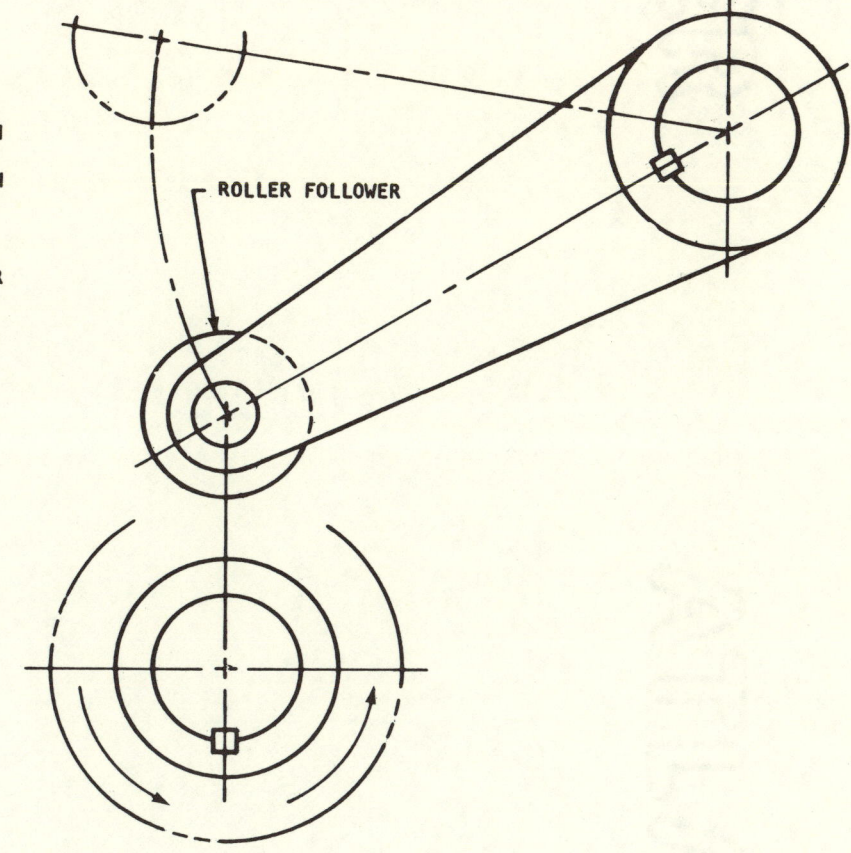

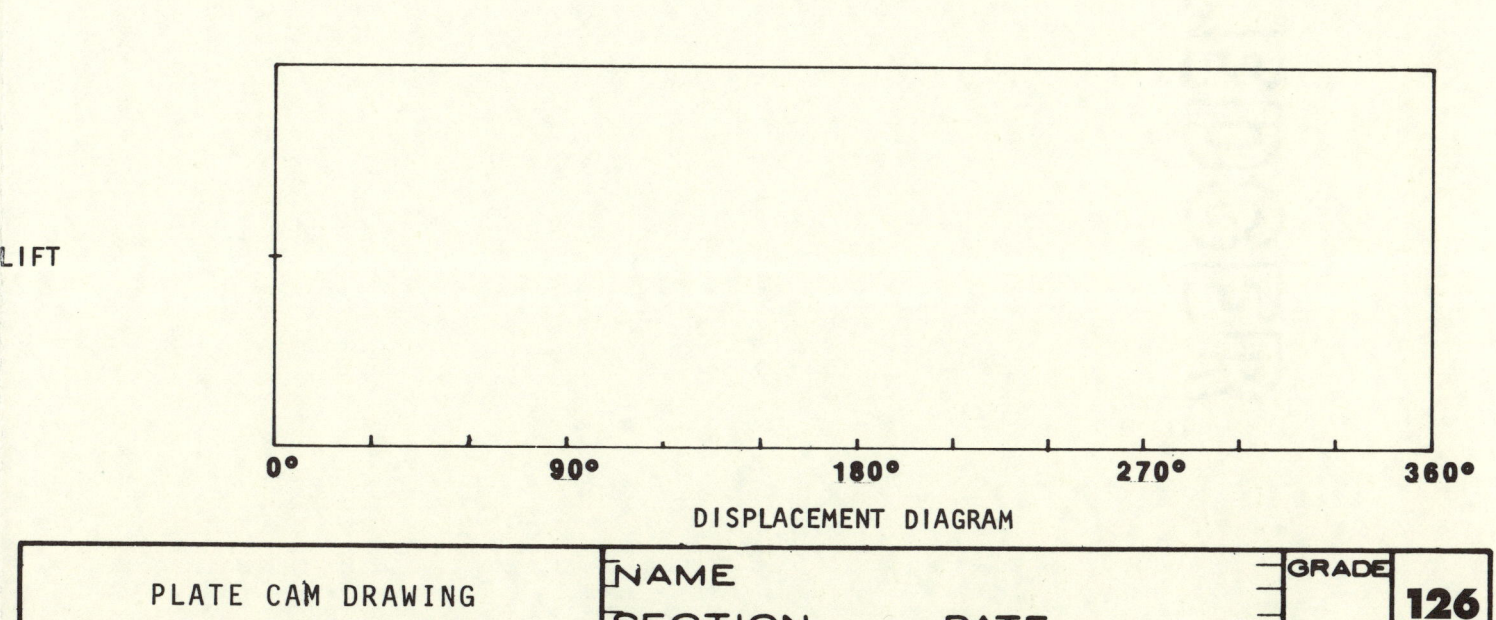

LIFT

0°   90°   180°   270°   360°

DISPLACEMENT DIAGRAM

PLATE CAM DRAWING | NAME | GRADE
SECTION   DATE | 126

PROVIDE WELD SYMBOLS FOR THE PROBLEMS DRAWN AND DESCRIBED BELOW.
SKETCH OR DRAW YOUR INTERPRETATION OF THE SYMBOLS ON THE DRAWINGS.

1 DRAW A WELD SYMBOL FOR A FILLET WELD

2 DRAW A WELD SYMBOL FOR A BUTT JOINT WITH SINGLE BEVEL GROOVE WELDS ON EACH SIDE OF ONE PLATE. BEVEL ANGLE 45°

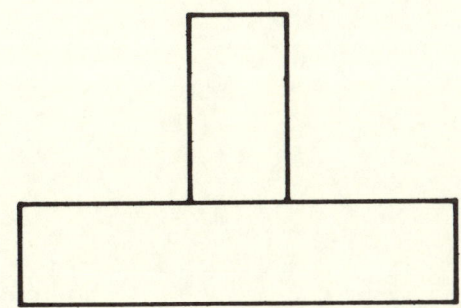
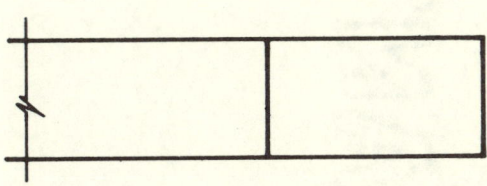

3 DRAW A FILLET WELD SYMBOL ON BOTH SIDES OF THE PART. DO NOT INCLUDE THE ANGULAR CUT EDGE.

4 DRAW A WELD SYMBOL FOR A COMBINED FILLET AND DOUBLE BEVEL GROOVE (60°) WELD. DEPTH OF GROOVE WELD 3/8".

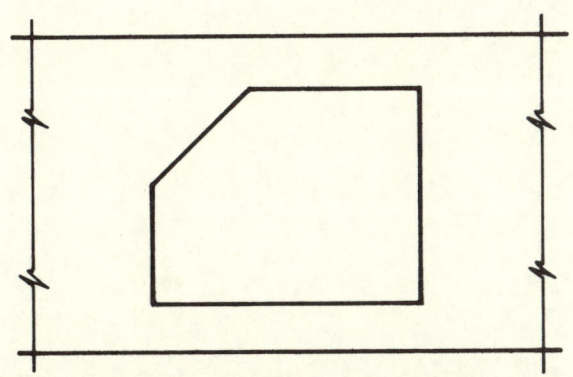
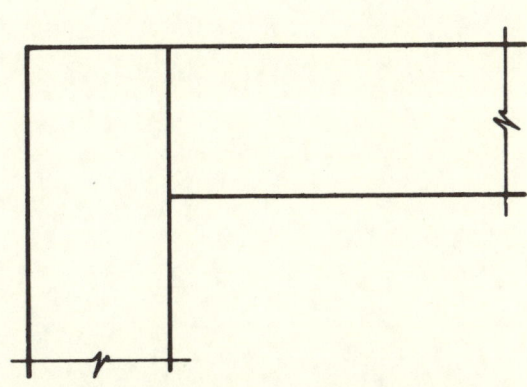

5 DRAW A WELD SYMBOL FOR A SQUARE GROOVE BUTT JOINT. THE ROOT OPENING IS 1/16". THE WELD PENETRATION IS 1/8".

6 DRAW A WELD SYMBOL TO REPRESENT A SINGLE OR MULTIPLE PASS WELD BUILD-UP OF THE SURFACE, 3/16" HIGH.

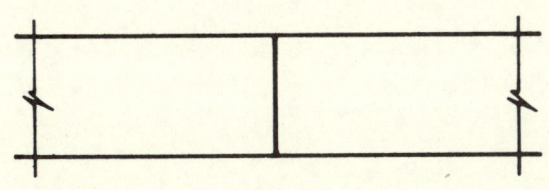

WELDING SYMBOLS

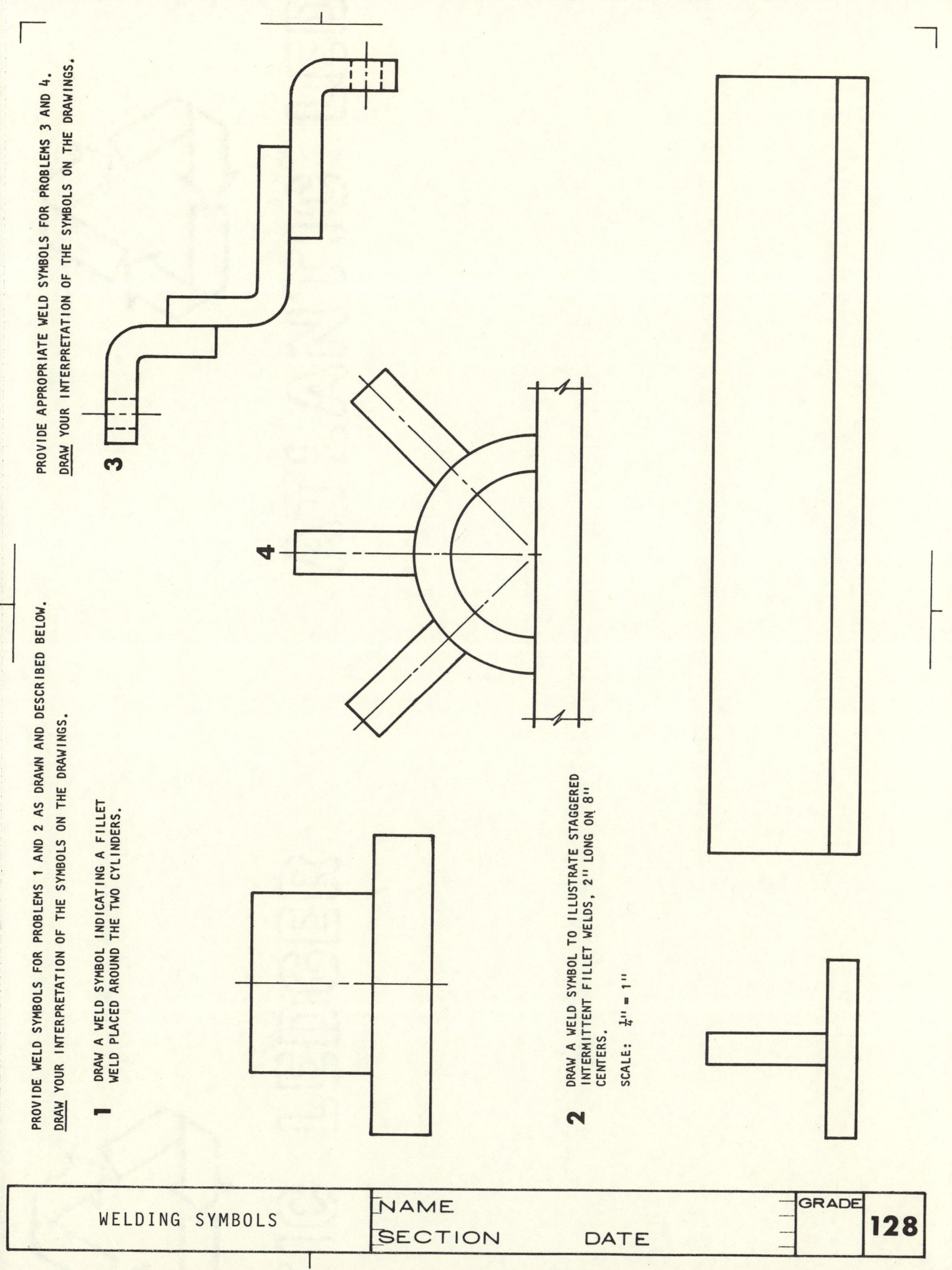

PREPARE AN INSTRUMENT WELDING DRAWING
OF THE LINK IN THE SPACE BELOW.

DIMENSION THE DRAWING COMPLETELY WITH
APPROPRIATE WELD SYMBOLS FOR EACH JOINT.

SCALE: EACH GRID UNIT ON THE PICTORIAL
EQUALS ONE QUARTER INCH.

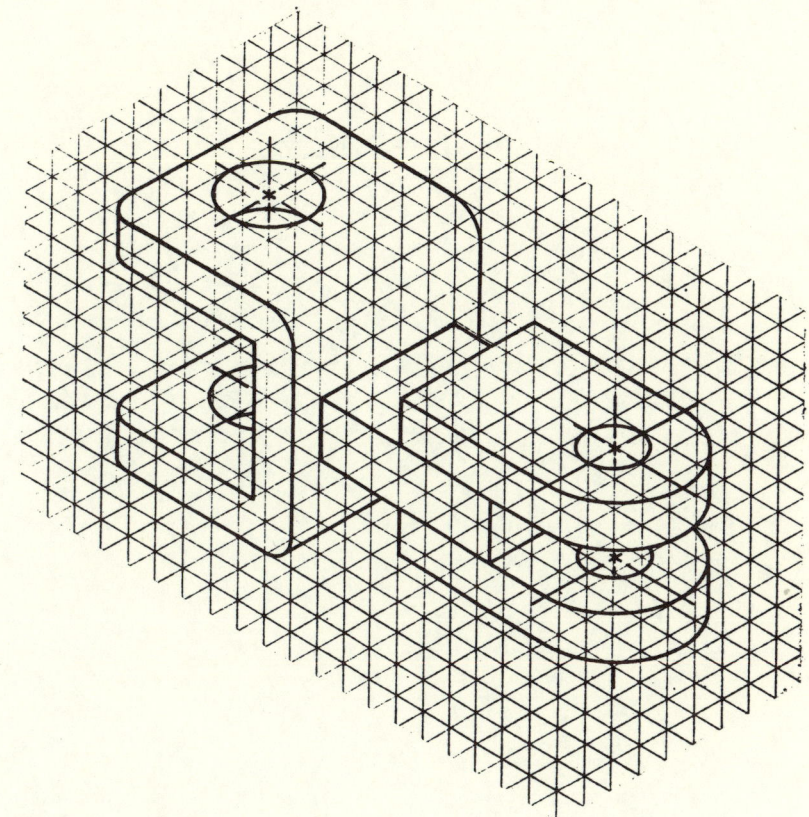

WELDING DRAWING

NAME

SECTION   DATE

GRADE

129

THE ELECTRONIC INTEGRATED CIRCUIT WIRING DIAGRAM ILLUSTRATED BELOW INCORPORATES A 555 TIMING CIRCUIT TO OPERATE A FLASHING LAMP.

DRAW THE FOLLOWING:

1. DRAW AN ENLARGED SCHEMATIC DIAGRAM OF THIS CIRCUIT IN THE SPACE BELOW.
2. WHEN TWO LAMPS ARE USED, ALTERNATELY FLASHING LIGHTS CAN BE INCORPORATED. PLACE A SECOND LAMP IN THE CIRCUIT FOR THIS OPERATION.
3. ADD A MICROSWITCH TO THE CIRCUIT TO PROVIDE MANUAL OPERATION.

INTEGRATED CIRCUIT TIMER

555

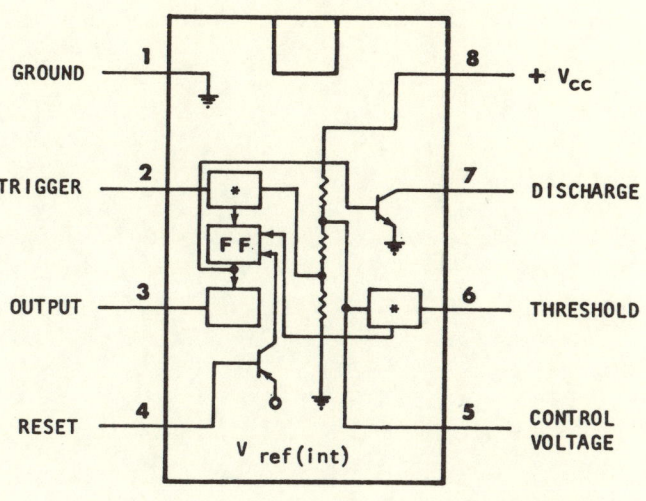

* COMPARATOR

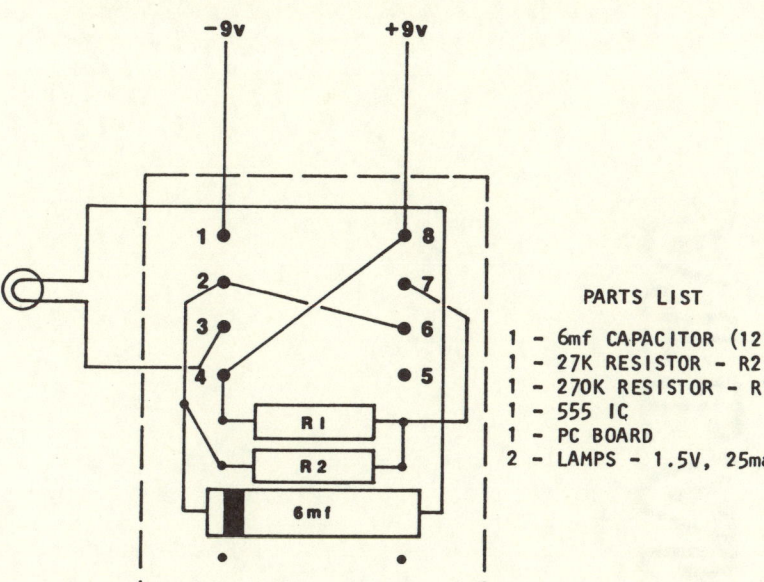

PARTS LIST

1 - 6mf CAPACITOR (12V)
1 - 27K RESISTOR - R2
1 - 270K RESISTOR - R1
1 - 555 IC
1 - PC BOARD
2 - LAMPS - 1.5V, 25ma.

FEATURES

NORMALLY ON OR NORMALLY OFF OUTPUT
TIMING FROM MICROSECONDS THROUGH HOURS
OPERATES IN EITHER ASTABLE OR MONOSTABLE MODES
ADJUSTABLE DUTY CYCLE
TEMPERATURE STABILITY BETTER THAN 0.005%/°C

ELECTRONIC DRAWING

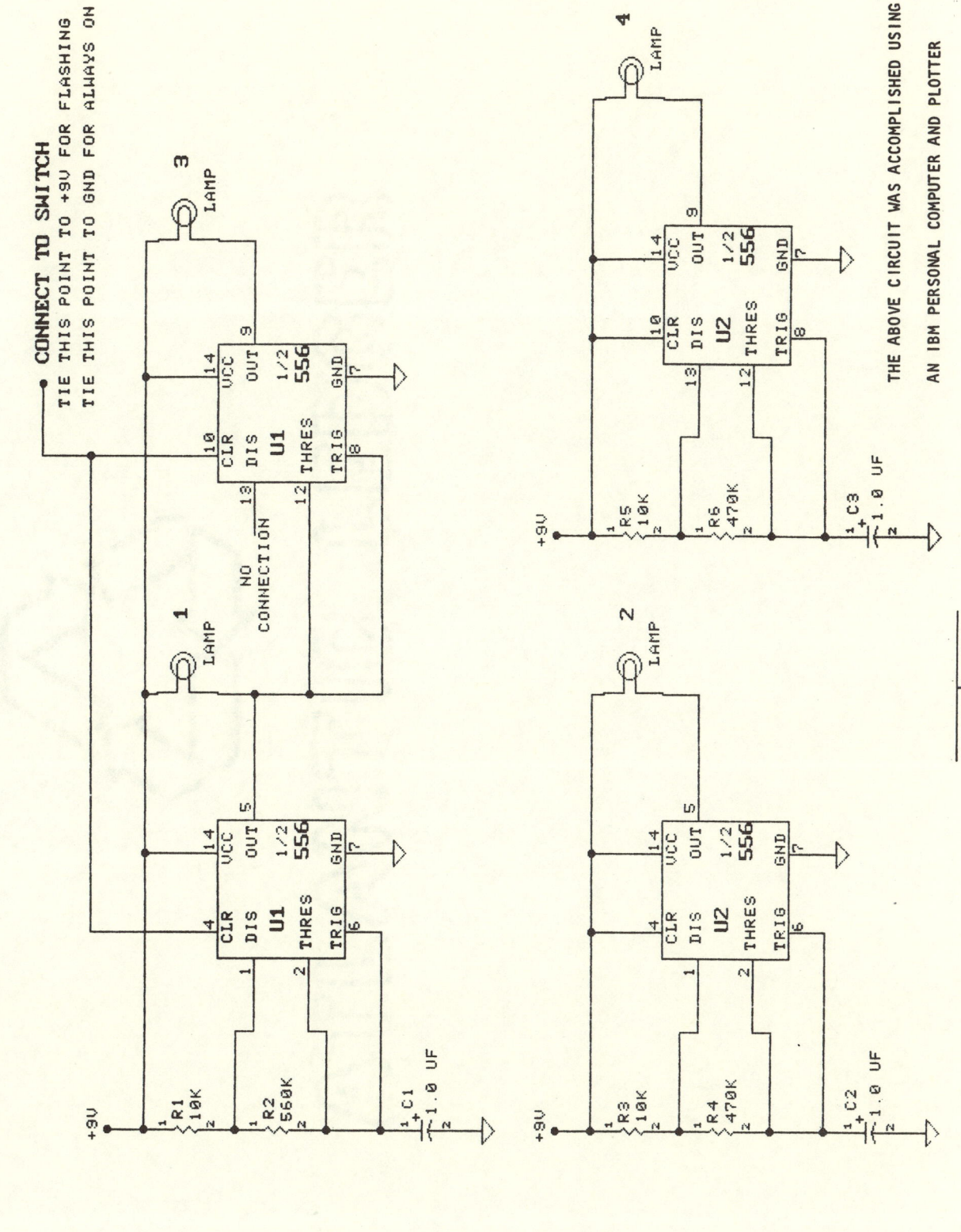

ELECTRONIC DRAWING

THE DRAWING AT THE RIGHT IS AN ISOMETRIC VIEW OF A SECTION OF A PIPING SYSTEM.

BEGIN WITH POINT "A" AND DRAW THE PLAN AND SOUTH ELEVATION OF THE PIPING SYSTEM USING THE SCALE INDICATED.

REFER TO YOUR TEXTBOOK AND SELECT THE PROPER FITTINGS TO CONNECT THE PIPE SECTIONS ACCORDING TO THE SPECIFICATIONS GIVEN BELOW:

1. SINGLE-LINE SYMBOLS, WELDED CONNECTIONS.
2. SINGLE-LINE SYMBOLS, FLANGED CONNECTIONS.
3. SINGLE-LINE SYMBOLS, SCREWED CONNECTIONS.

SCALE: ¼" = 1'- 0"

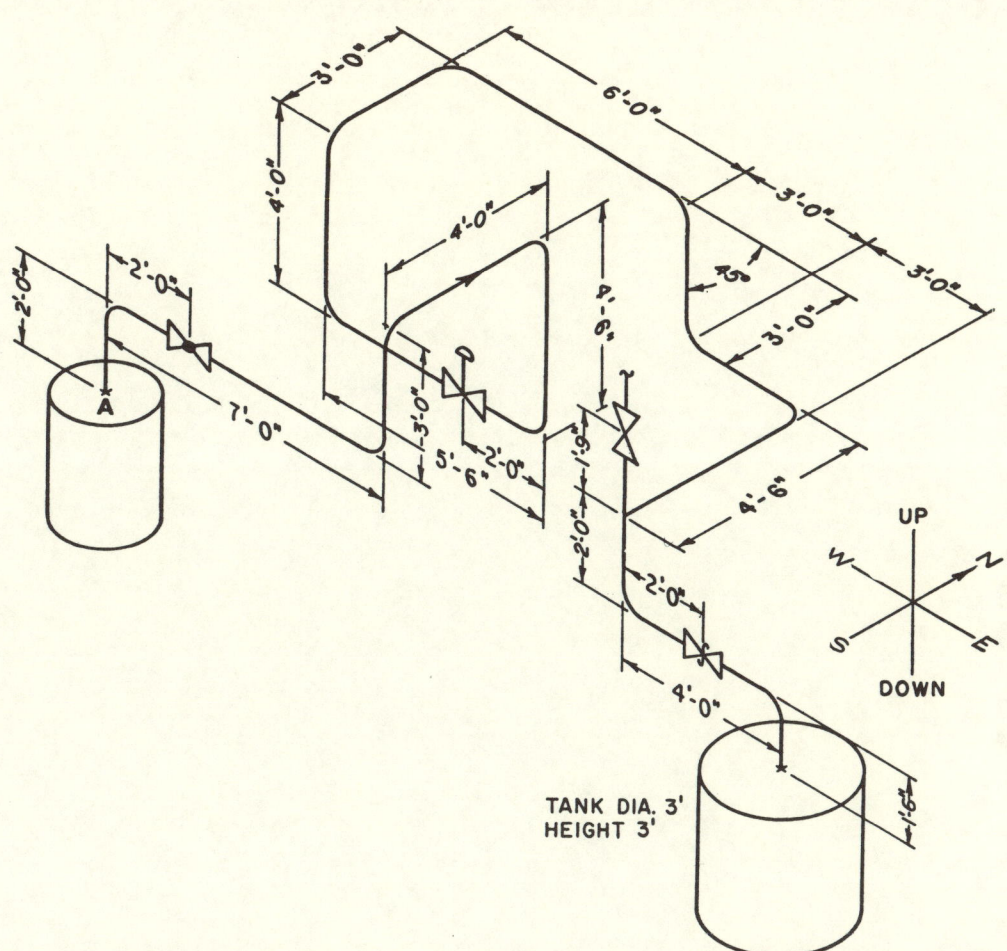

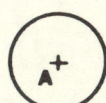

PIPE DRAWING

132

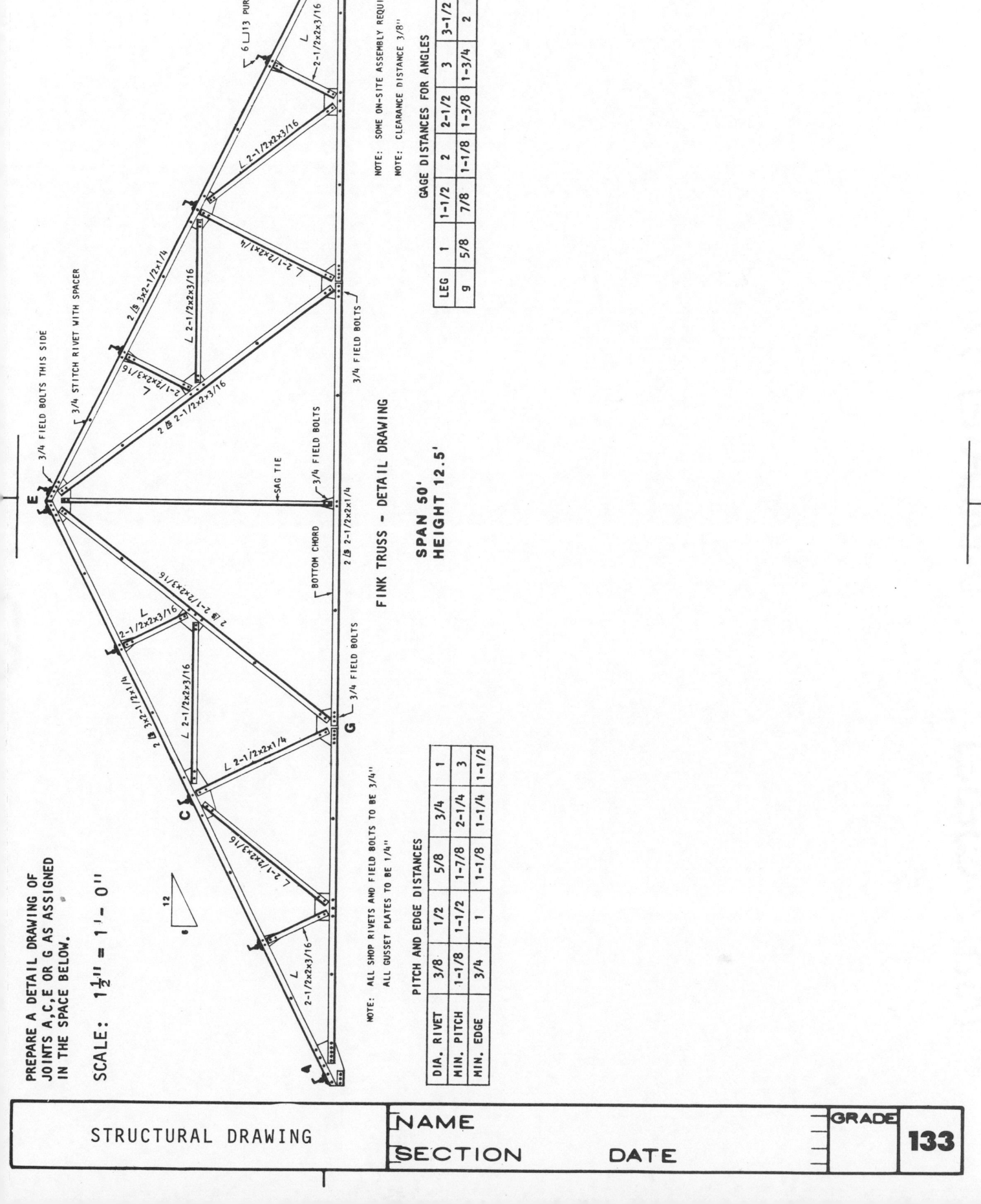

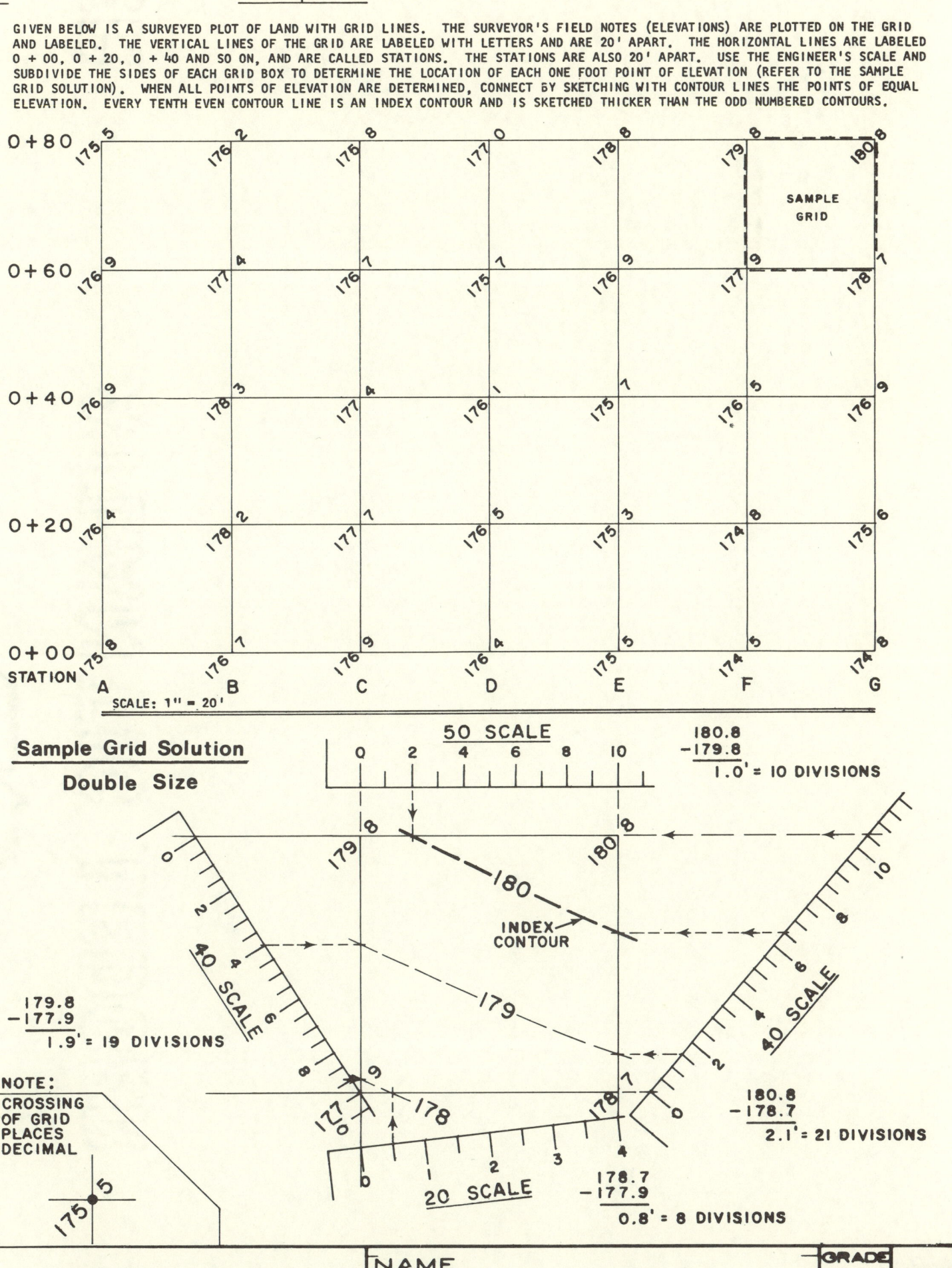